Alexander Unzicker

THE HIGGS FAKE

How Particle Physicists Fooled the Nobel Committee

Alexander Unzicker

The Higgs Fake

Copyright © 2013 Alexander Unzicker
All rights reserved.
ISBN: 1492176249
ISBN-13: 978-1492176244

Contents

Preface
Why This Had To Be Said   5

Part I
The Sicknesses of Particle Physics   11
Chapter 1
Overwhelming Complication
Why Good Physics Needs To Be Simple   11
Chapter 2
Suppression of Fundamental Problems
How the Important Questions Fell into Oblivion   20
Chapter 3
Historical Ignorance
Why Particle Physics Continues To Repeat the Old Errors   31
Chapter 4
Evidence Shmevidence
Endlessly Refined Sieving: How Particle Physicists Fool Themselves   40
Chapter 5
The Big Parroting
How Incestuous Expert Opinions Spread   52
Chapter 6
Lack of Transparency
Why Nobody Can Oversee That Business   62

Part II
How It Came All About   73
Chapter 7
The Steady Degradation of Thought
From Poorly Understood Physics to High-tech Sports   73
Chapter 8
The Takeoff to Metaphor
Brute Force Finding, Brute Force Explaining   83
Chapter 9
Approaching Fantasy Land
Halfway between Science and Non-Science   95
Chapter 10
Towards the Summit of Absurdity
Arbitrary Theories and Banal Facts   105
Chapter 11
The Higgs Mass Hysteria
If Anything, the Hype of the Century   115

Part III
Antidotes   125
Chapter 12
Who Is Telling and Selling the Nonsense
The Big Show of Boasters, Parroters and Fake Experts   125
Chapter 13
Against Belief: What We Need to Do
...And What You Should Ask These Guys   134
Literature   147

Alexander Unzicker

PREFACE

WHY THIS HAD TO BE SAID

The 2013 Nobel Prize in physics was awarded very soon after the announcement of the discovery of a new particle at a press conference at CERN on July 4, 2012. The breaking news caused excitement worldwide. Yet the message conveyed to the public, as if something had happened like finding a gemstone among pebbles, is, if we take a sober look at the facts, at best an abuse of language, at worst, a lie.

What had been found by the researchers did not resolve a single one of the fundamental problems of physics, yet it was immediately declared the discovery of the century. Whether this claim is fraudulent, charlatanry, or just thoroughly foolish, we may leave aside; that the greatest physicists such as Einstein, Dirac or Schrödinger would have considered the "discovery" of the Higgs particle ridiculous, is sure. They would never have believed such a complicated model with dozens of unexplained parameters to reflect anything fundamental. Though on July 4, 2012, the absurdity of high energy physics reached its culmination, its folly had begun much earlier.

I shall argue that particle physics, as practiced since 1930, is a futile enterprise in its entirety. Indeed physics, after the groundbreaking findings at the beginning of the twentieth century, has undergone a paradigmatic change that has turned it into another science, or better, a high-tech sport, that has little to do with the laws of Nature. It is not uncommon in history for researchers to follow long dead ends, such as geocentric astronomy or the overlooking of the continental drift. Often, the seemingly necessary solutions to problems, after decades of piling assumptions on top of each other, gradually turn into something that is ludicrous from a sober perspective. A few authors, such as Andrew Pickering and David Lindley, have lucidly pointed out the shortcomings, failures and contradictions in particle physics in much detail, providing, between the lines, a devastating picture. Though their conclusions may not be very different from mine, I cannot take the detached perspective of a science historian. It annoys me too much to see another generation of physicists deterred by the dumb, messy patchwork called the standard model of particle physics that hides the basic problems physics ought to deal with.

Therefore, I shall be very explicit in this book. It is written for the young scholar who wants to dig into the big questions of physics, rather than dealing with a blend of mythology and technology. It should demonstrate to the majority of reasonable physicists that the high energy subsidiary is something they would be better getting rid of, because it doesn't meet their standards. All scientists who maintain a healthy skepticism towards their particle colleagues should be encouraged to express their doubts, and the general public, many of whom intuitively felt that the irrational exuberance of July 4, 2012 had little to do with genuine science, should come to know the facts. Last but not least, it should provide journalists and people responsible for funding decisions with information they need to challenge the omnipresent propaganda.

To Whom It May Not Concern

Beware of false knowledge, it is more dangerous than ignorance – *George Bernard Shaw*

Needless to say, this book will hardly appeal to particle physicists, and not even lay much of a basis they will wish to discuss. There is no way to convince an expert that he or she has done nonsense for thirty years. Over the decades, high energy physicists have been hunting for ever rarer effects, just to declare as new particles everything they did not understand. Their model has grown to a nonsensical complexity nobody can oversee, thus their convictions about it rely –

much more than in any other field of physics – on trust in expert opinions (one might even say parroting). As a consequence, in any discussion with particle physicists one soon comes to know that everything is done properly and checked by many people. If you still express slight doubts about the complication, they will easily turn stroppy and claim that unless you study their byzantine model thoroughly, you are not qualified to have an opinion. But you don't have to be an ichthyologist to know when a fish stinks.

An obvious argument to make is that more than 10,000 physicists, obviously skilled and smart people, would not deal with a theoretical model if it was baloney, and presumably this is the strongest unconscious argument for all of them. It is a flawed argument, however, disproved many times in history. And it is inherently biased because it disregards all other physicists (probably the majority) who intuitively realized at the outset of their careers that a giant experiment involving a huge number of people was not the field where their creative ideas would flourish. Quantum optics, astrophysics and fields like nanotechnology have attracted the most talented in the past decades. No one who had a proper appreciation for the convictions of Einstein, Dirac, Schrödinger, Heisenberg or de Broglie could find satisfaction in post-war particle physics. This does not mean that all high energy physicists are twerps. Religion is said to make good people do evil things. To make intelligent people do stupid things, it takes particle physics. Many scientists, by the way, are busy fighting the religious nonsense that pervades the world's societies (let some political parties go unnamed). Intellectually, this is a cheap battle, and thus some are blind to the parallels of science and religion: groupthink, relying on authority, and trust to the extent of gullibility.

Though people will accuse me of promoting a conspiracy theory, I deny the charge. Most high energy physicists indeed believe that what they are doing makes sense, but they are unable to disentangle their belief from what they think is evidence. The more thoroughly one examines that evidence, however, the more frail it becomes. But, above all, it is impenetrable. Only the super-specialized understand their small portion of the data analysis, while a superficial babble is delivered to the public. This is a scandal. It is *their* business, not anyone else's, to provide a transparent, publicly reproducible kind of evidence that deserves the name. It is no excuse that, unfortunately, there are other degenerations of the scientific method in the realm of theoretical physics: supersymmetry, and string theory which never predicted anything about anything and never will.[1] It is a sign of the rottenness of particle physics that nobody has the guts to declare the nontestable as nonsense, though many know perfectly well that it is. They are all afraid of the collateral damage to their own shaky building, should the string bubble collapse. The continuous flow of public funding they depend so much on requires consensus and appeasement. However, experimental particle physics is somehow more dangerous to science as a whole, because with its observational fig leaves, it continues to beguile everybody that they are doing science instead of just pushing technology to the limits.

I don't care too much about the public money being wasted. We live in a rotten world where billions of dollars are squandered on bank bailouts, while every ten seconds a child dies of hunger. But the worst thing about the standard model of particle physics is the stalling in the intellectual progress of humankind it has caused. We need to get rid of that junk to evolve further.

How To Read This Book

> *My aim is: to teach you to pass from a piece of disguised nonsense to something that is patent nonsense.* – Ludwig Wittgenstein

The first part of the book will address the principal sicknesses of particle physics, starting with a discussion of the excessive use of arbitrary, unexplained numbers in the standard model, which has always been a sign of morbidity of a theory. None of the fundamental questions that bothered the founding fathers of the successful physics of the early twentieth century are solved today, most of them being drowned out by the current drivel. A look at history reveals that there is plenty of indication that particle physics has long been

---

[1] I addressed that in my book *Bankrupting Physics*.

heading down a dead end, a crisis characterized by anomalies and adhoc fixes as described by the philosopher Thomas Kuhn.

As high energy physics is practiced today, there is plenty of room to suspect that many of its results are instrumental artifacts due to extensive filtering or theoretical mismodeling. However, even if one is confident of the analysis, it is easy to see that particle physicists continuously declare ever rarer, though banal, effects to be manifestations of their wishful theoretical thinking. By construction, it is a seemingly never-ending, epistemologically absurd process, supported by its slowness that hides the sociological nature of opinion creation by groupthink.

As happened with the Higgs, several times something has been declared as the final missing piece of the standard model. Therefore I shall tell the story of high energy physics with a somehow blasphemous point of view in part II. It is for the reader who wants to see the evolution of particle physics – how, within a few generations, the field turned its paradigms upside down and slipped into absurdity. The busy reader may skip part II and jump to part III which contains a description of the most famous popularizers and the nonsense they distribute all over the world. Luckily, YouTube is a unique archive of their ridiculous claims.

I shall also propose a way to get out of the crisis. What is desperately needed is a new scientific culture of transparency, a world in which results can be tested and repeated step-by-step by an open, unrestricted community of researchers and publicly available data. In the final chapter, I give a list of questions I would love to hear asked at press conferences, hearings, and conferences. Sometimes it would be so easy to unmask the superficial verbiage with which the field continues to throw dust in the eyes of everybody. Revenue in the form of public recognition is highly desired, and CERN lobbyists have been working hard to get it.

One particular dust-throwing event, and the inspiration for this book, was the Nobel Symposium on the Large Hadron Collider results in May 2013 in Stockholm. Coming from different fields, the members of the Nobel Prize committee have to arrive at an agreement – which is usually accomplished by a quota of awards to the various fields. Currently, there is one string theorist, one dark matter/SUSY enthusiast, and one high energy physicist (all of them giving talks at the above seminar) on the committee, plus three reasonable physicists who, after the global hype had been generated in 2012, eventually caved in. The Royal Academy, then, had little choice but to meet the expectations. Particle physics, once again, had succeeded in selling its hokum. Therefore, it is time to stop seeing the Nobel Prize as a sacrosanct accolade for physics. In the course of the last 50 years, the award has contributed considerably to the degeneration of the search for the fundamental laws of Nature.

PART I

THE SICKNESSES OF PARTICLE PHYSICS

CHAPTER 1

OVERWHELMING COMPLICATION

WHY GOOD PHYSICS NEEDS TO BE SIMPLE

Democritus wasn't stupid. His idea of atoms being the elementary building blocks of Nature was essentially a simple concept. Though he did not have a modern picture of the atom, he was convinced that such simplicity was necessary in order to understand Nature. Our current understanding in the twenty-first century instead goes as follows: Four different interactions exist in Nature, and the building blocks of the atomic nuclei, protons and neutrons, consist of parts themselves: quarks and gluons. The quarks show up in six distinct species called "flavors," each of which can appear in three different "colors." Besides heavy particles there are two other groups, middle-weights and a light sort, consisting of electrons, muons, tauons, and three, if not more, corresponding types of neutrinos, not to forget W, Z and Higgs bosons. All these particles not only carry mass and elec-

The Higgs Fake

trical charge, but also so-called "isospin," "charm," "bottomness" and another couple of characteristics which define them. It would fill several pages if I tried to give you just a rough idea of the underlying notions. But let's pause for a moment. Imagine you are visiting another civilization where a shaman tells you the above story. Suppose you were never taught modern physics, would that seem convincing to you? Or rather a fancy piece of mythology? But not only that. It's not the uneducated are too dumb to appreciate such sophisticated concepts, the greatest minds to which we owe our civilization would be super-skeptical of that story. Einstein didn't refer just to his period when he said that laws of Nature are unlikely to be true unless they are *very* simple. (A stupid objection is "things are not as simple as they were then" – like an obese adult saying "I am not as slim as a child".) Einstein and his contemporaries disdained any theory with unexplained numbers – nowadays dozens of them are needed to describe the above multitude of particles. Dirac worried about *two* elementary particles being a suspiciously complicated model, and Newton's supreme credo was simplicity. At the mere sight of today's model, they would all get *physically* sick.

Occam's Stylists

Did Newton, Maxwell, Einstein, Schrödinger, Dirac all share a near religious predilection for simplicity that – sorry, sorry – turned out to be wrong? No. It was their very experience that their accomplishments consisted of simplification. Once they had understood, the complicated mess went away. Thus there is actually pretty good *evidence* that simpler means better. Ironically, this is something literally every modern physicist would agree on, and it goes back to the medieval English philosopher William Occam who stated that, among concurring theories, you should take the simpler one. The metaphorical expression "Occam's razor" refers to a sharp blade to which complicated models fall victim. Despite accepting the methodological virtue of Occam's razor, particle physicists have avoided being shaved for half a century because they think it's not for them. But believing that relative simplification is good while absolute simplicity is unnecessary is schizophrenic. They just don't realize how absolutely unlikely their messy model is.

Dazzling with Probabilities

On the other hand, scientists give their discoveries, such as the Higgs, extremely high probabilities, claiming that it is more than 99.9999 percent assured, or, in the jargon, "five sigma," which refers to Gaussian statistics. But this is just throwing a smokescreen on the results. It is not the *statistical* fluctuations one has to be concerned about, but the risk of systematic errors − faulty assumptions, computer bugs, material effects, but also self-deception − which are never as rare as the Gaussian modelers would like them to be. The same type of flawed analysis, conducted by blinkered specialists, has been identified in economics by Nassim Taleb. In his book *The Black Swan* Taleb demonstrates that underestimating low probabilities, combined with sufficient gullibility with respect to the theory, is responsible for the catastrophic crashes we observe once in a while. And in both cases it is the excessive complication of the system that makes modeling errors more likely.

Above all, however, all the impressive probabilities are valid under the assumption that the model is correct. There is an incredible number of hypotheses layered one over another, and even if the probability of a single one being true is fairly high, the whole edifice is a fragile thing. It cannot be tested in its entirety any more, even if the experts claim to "know it for sure." But no sigma whatsoever can tell them if the whole model is bullshit.

Experiment Ltd

As I said, simplicity is empirically tested by the history of science, but it is also a concept of which the greatest thinkers of humankind were convinced. Not to bother at all with the issue of simplicity in the laws of Nature, as particle physicists do, requires a certain philosophical illiteracy.

The Higgs Fake

Philosophy, by the way, is not just empty verbiage (as some arrogant physicists deem it) but means the desire to know who we are and where we come from. Natural philosophy is the origin of physics. And even if most of today's philosophers may be pretty ignorant about physics, it is not a good idea for physics completely to forget its parent corporation. The extraordinary success that made physics evolve from pure philosophy was grounded on the experimental method (though high energy physicists have steadily perverted the notion of experimental evidence), the evidence-based reasoning that caused the scientific revolutions in the past centuries. And, by the way, we really need to require testability for debunking fantasies like strings, parallel universes or "chaotic inflation". The experiment-based collection of data is not a virtue in itself. The development of fundamental science on a larger scale calls for a philosophical oversight of the industrial production of empirical facts. Physics accumulates facts like a prospering economy produces goods. But, be it the flood of consumer goods or the data stream flowing out from CERN, you have to ask what the hell you need it for. Without the philosophical aspiration to understand the fundamental constructions of Nature, physics degenerates like a civilization without values. And occasional speculation bubbles without real productivity have already led branches of theoretical physics into bankruptcy.[1]

Let's Listen to the Thinkers and to Music

Besides the principle of simplicity inherent in natural philosophy, we need a few more rules to do good science. A Rhadamanthine tool to sift the chaff of fantasy from the wheat of research is the *Logic of Scientific Discovery* by Karl Popper. Popper's criterion of falsifiability is an effective antidote to any esoteric hokum: Be falsifiable. A hypothesis must predict something and leave the possibility of being *rejected* by observation, if it is scientific fair-play and not ideology. Predicting *anything*, however, is not a great merit, you have to come out with a measurable number in order to be credible – keep that in mind for

---

[1] See my book *Bankrupting Physics* with Sheilla Jones (Palgrave 2012).

our later discussion. Again it was Isaac Newton who, in the lingo of the day, put it straight and simple: "God created everything with number, weight and measure." I mention this because it bluntly summarizes the methodological failure at the heart of particle physics. In short, the disrespect of Newton's statement is the reason why, instead of getting hard facts from particle physics, we are swamped by so much semi-metaphorical shit.

Allow me to mention a few caveats about the notion of simplicity, since it has also been abused by theoretical fantasizers dealing with their "beautiful" or "natural" symmetry groups in higher dimensions (e.g. the aesthetically challenged babble about the beauty of "supersymmetry"). When we talk about simplicity in scientific theories, it is helpful to have a clear definition at hand. It is *not* beauty, it is *not* condensing something into short mathematical formulas, it is just one thing: reduce the information. Mozart is said to have jotted down a piece of music after listening to it, though memorizing the stream of data would have been impossible for purely physiological reasons. However, he didn't hear random notes but a structure that rigidly followed the laws of harmony. Mozart *understood* those and was able to reduce the thing to a few distinct facts. Ancient observers of the starry sky must have been overwhelmed by the sensory stream like those who listen to a piece of music. But Nature follows a simple law: all planetary motion is determined by the product of the sun's mass and the gravitational constant, called Kepler's constant after the discoverer of the (elliptical) orbits. It is rare that a discovery has reduced the information so dramatically: the data of thousands of single observations, a seemingly complicated picture, boiled down to a few numbers, once the system was understood. Scientific insight, in short, is organizing and reducing the information gained by observing.

The Breakthrough Mechanism

And that's how true science has *always* worked. Maxwell's theory and the discovery of electromagnetic waves in 1888, which has revolutionized our life, is condensed in the formula $\varepsilon_0 \mu_0 = 1/c^2$, which relates the electromagnetic constants to the speed of light c. It simply reduces the number of independent constants from three to two. *This* was the breakthrough of a century, not the claptrap at a certain press conference on July 4, 2012. Max Planck's law of thermal radiation, published in 1900, condensed two separate rules into one, dominated by the newly found action quantum h, maybe the most important constant of Nature. Einstein used it to explain in a simple way the phenomenon of light quanta with the energy $E = hf$ (f being the frequency), and Niels Bohr, another visionary of the early twentieth century, deduced hundreds of spectral lines of the hydrogen atom with h.[I] Einstein, who also reflected on how these revolutions came about, contended that a complete physical theory should not contain any number which comes along unexplained.[1] It's just a deficit of understanding.

We may be far from such a complete theory, but it is as clear as day that the fewer numbers you need, the better your theory is. If you find rigorous rules and a few important constants, you are closer to Nature's secrets, if you need the same amount of parameters as your data already had, you are treading water. Particle physics is treading water, and not in a pool but a giant swamp. Having lost the overview, they no longer even count their unexplained numbers. A conservative estimate is 18, but that neglects the masses and other properties of the heap of particles they're dealing with. Go and ask a high energy physicist about the number of free parameters in the standard model, he will quibble.[II] It's just a mess. It upsets me to hear the above epochal discoveries named on an equal footing with the baloney that comes out of the big collaborations who, despite their computers, cannot count their parameters any more. Though these considerations seem pretty evident, let me mention some counterarguments, loosely sorted by stupidity.

[I] Using a formula previously discovered by Johann Jakob Balmer – another considerable simplification.

[II] I can't resist reporting the following. One particle physicist, in all seriousness, said the standard model had just three (!) parameters, later specifying that he intended the "electroweak boson sector". It's like a drunk saying, "I drink very little alcohol in the champagne sector."

Alexander Unzicker

Dumbfounding Answers

There is a species of high energy physicist who lives happily in the clutter. They seem to regard their experiments (which excrete one number after another) as a botanic garden with its variety of forms and colors that prompt us to admire how wonderfully multifaceted Nature is. These people can be diligent observers, and if they're happy with their world view, well, all right. But if you do science, which means using your cerebrum to process the data, it's your damn job to understand. It might be difficult, yes, but if you don't even try, please get out of this business.

Then, there are people who don't have any problem because they think they are working with a valid model. One particle physicist, in all seriousness, compared the unexplained numbers of the standard model to the masses of planets, saying: "Well, Newton didn't predict these numbers either, did he? So what?" I just can't get it into my head how one can compare a random conglomeration of dust in the solar system to the fundamental characteristics of particles present in the whole universe. I hope that Newton was never confronted with such a silly demand, but that's not the bottom yet.

One particle physicist, educated at Hamburg's DESY laboratory, said to me: "Look at the building across the street. You surely would need a lot of parameters to describe it precisely. So what problem do you have that we use many parameters for the system of elementary particles?" My problem was I was slack-jawed, because I had never (including experience with clueless students and Jehovah's Witnesses) heard such a dumb argument. This guy had a PhD! But with all the busy bean counting he had done in his lab life, it had never crossed his mind that intricately ornamented architecture is something slightly different from a law of Nature.

Being Out to Lunch

Now we come to the more severe cases who deny the complication. More than fifty arbitrary numbers is still simple, right? Mhm. "Consider that," they claim hastily, "the standard model was a tremendous simplification with respect to the several hundred elementary particles detected in the 1960s!" When particle physicists talk about history, their memory is limited to about ten years, and you may stumble when denying the argument, as happens when you are confronted with people thinking in their box. Fifty numbers are less than several hundred, yes, but who the hell told you to produce such a bulk of particles you don't understand? Please tell me who believed that this crap of *several hundred* particles had any fundamental significance? Used to a daily case of wine and now you are happy with a bottle? Congratulations. Such an attitude is the prototypical reasoning of people who, to quote Oscar Wilde, know the price of everything but the value of nothing.

Another way to deny the complication, preferred by theorists, is to just fade out the nasty amount of numbers and contemplate the wonderful "underlying structure" of symmetry groups. Such a disconnect from reality surely makes it easier to view things as simple. In a debate I had with a CERN theorist, my opponent insisted that the symmetry group of the standard model $U(1) \times SU(2) \times SO(3)$ had "just eight numbers." Well, this sum of group dimensions has nothing to do with anything real, but it certainly feels better to count just the number of garbage cans rather than to sort the content.

> *The simplification of anything is always sensational.* – G. K. Chesterton, English writer

Split-Brain Expertise

Now we come close to madness. There are folks among the particle physicists who admit that the standard model may be no better than the medieval geocentric model of the solar system, but it is the best (some say, the only) model we have, so let's continue to work with it. Yeah! Surely Newton would have had no chance to find out the true law of gravitation if, instead of Copernicus and Kepler who doubted the old model, such morons had been at work. And particle physicists still behave in such a manner, after history has taught us that lesson! I don't understand, but it appears that humans have an inherent stubbornness which makes them continue their habits even after logic has told them it's senseless. Nassim Taleb, while describing a similar "it's the only model we have"-attitude in economics, asks: Would you fly with a pilot across the Alps who uses a map of the Pyrenees, justifying himself with "it is the only map we have"? It is certainly easier to measure than to know what you are measuring. And some (particularly adherents of supersymmetry) cherish the illusion that further experiments (in the same thoughtless manner) *might* give us *clues* for where to search for *new physics* beyond the standard model. They advocate muddling along in order to get ideas for a cleanup, but expecting the simplifying discovery on the same track that has complicated things for decades is mad, comparable to a hiker expecting a mountaintop with panoramic views suddenly to appear after having descended into a deep canyon. Or, as the comedian Will Rogers put it: "If stupidity got us into this mess, why can't it get us out?"

And finally (I may be boring you by quoting dull statements, but that's what you happen to hear), there are people who entered science accidentally when they would have been better off doing some esoteric hokum. They aren't disturbed in the slightest by the absurd complication of the standard model and hold the credo "Well, all this is so wonderfully complicated, who knows if humans will ever be able to grasp a meaning behind? Isn't it arrogant to assume we can understand Nature? Shouldn't we be more humble?" Such a statement is in the first place to admit one's intellectual impotence. Sure, nobody can prove that there are laws of Nature we can understand. But insinuating that Nature is such a jumbled mess instead of daring to consider that it's you who is just too dumb to understand, isn't a humble attitude after all.

To summarize, dear reader, I hope I have convinced you that simplicity means understanding, and this argument alone is a vital blow to the standard model. Nevertheless, I shall outline further, equally absurd aspects of that enterprise called particle physics.

CHAPTER 2

SUPPRESSION OF FUNDAMENTAL PROBLEMS

HOW THE IMPORTANT QUESTIONS FELL INTO OBLIVION

Thinking is the most unhealthy thing in the world. – Oscar Wilde

It's not exactly surprising that few problems of physics that were discussed in the 1930s are hot topics today. You may guess that most of them just have been understood. Not so. Instead, not a single one of the great riddles that bothered Bohr, Pauli and Schrödinger, the puzzles that Einstein and Dirac worried about until their last days, have been solved since then. I start my discussion with one problem that, while being a theoretical issue in the first place, has fatal consequences for modern accelerator physics. It screws up the analysis of particle interactions in modern collider experiments.

Since the discovery of electromagnetic waves in 1888 we know that accelerated electric charges radiate energy in space – that's what happens in every cell phone antenna. You might take for granted now that physics has a formula at hand that allows you to calculate the amount, wavelength and direction of the radiation[I] once you are given the acceleration of a charge in every moment. It hasn't. Despite a bunch of fixes, patches and approximations, there is no formula that is generally valid at strong accelerations. That means that when the highest (negative) accelerations produced by humankind occur, when protons[II] crash into each other at the Large Hadron Collider, nobody knows how to calculate the resulting gamma radiation precisely. There is no universal formula, no method, no theory. The simple reason is that electrodynamics, while being a widely successful theory, is incomplete when it comes to strong electric fields. This is not an ungrounded suspicion of mine, the theory itself is contradictory – an ugly fact that you can find in textbooks. Coulomb's law of electrostatics would predict that the electric field in the vicinity of an electron/proton becomes infinitely great – but this cannot be true because the poor electron would carry an infinite amount of energy, and according to Einstein's $E=mc^2$, an infinite mass. As Lev Landau in his treatise on theoretical physics points out,[2] that infinite basin of energy would allow a charge to radiate whatever amount in space – if the law were true. It cannot be. (Most particle physicists readily deny this consequence. Sometimes it is so easy to identify the ignorant.) Those high energy physicists who consider themselves quick-minded will respond that this old staff of classical electrodynamics is obsolete and superseded by the theory of quantum electrodynamics, which somehow (they would say they can't explain precisely how) fixes the problem. But that's rubbish. Richard Feynman, who got the Nobel for having developed quantum electrodynamics, writes in his Lectures: "The difficulties persist, even when electrodynamics is unified with quantum mechanics."[3] The psychological repression of this unpleasant fact also persists, unfortunately, among the entire community of data jugglers. Most are just wide-eyed or say in reflex manner that synchrotron radiation is well understood (has nothing to do with the collision accelerations), they don't even know the best argument they could come up – formula 14.14 in *Jackson's Electrodynam-*

[I] Technically, I(th, ph, lam,t) dependent on a(t).
[II] That applies even more to the earlier e^+e^- colliders.

ics. But unfortunately, that doesn't cover the general case. Thus there is not even a theoretical upper limit to such radiation losses, though such an assumption is usually made to soothe one's nerves.

> *We will be considered the generation that left behind unsolved such essential problems as the electron self-energy.* – *Wolfgang Pauli, 1945 Nobel Laureate*

A Massive Problem and Talking the Way Out of It

The underlying problem of our inability to calculate the electromagnetic energy of charged particles, and consequently, their mass, is a very basic one and bears consequences far beyond the collider experiments. The reason why the standard model cannot calculate mass is stunningly simple.

Whatever formula may calculate the mass of a particle, it has to have the physical unit of a mass, which is kilograms, and those kilograms have to come out of the formula, you can't pull it out of your ass. The only way to do it is to find combinations of the fundamental constants of Nature, otherwise you are cheating. But there is *no* such combination, as every high school kid can understand.[I] I'll pay you 1000 dollars right away if you find a combination that does. There is just one caveat – you can't do it unless you use the gravitational constant G, that is, you consider gravity in your model. But particle physicists don't do it because gravity is a difficult business that goes above their minds. It's outright lunatic that they declare that calculating masses, alas, is impossible in the standard model[II] and at the same time deliberately disregard gravity which would offer the only possibility to try it.

Since Isaac Newton, mass is the important, the basic, the essential, the most intriguing property in fundamental physics (which by the way would allow clean and testable predictions). It's a scandal that the standard model talks its way out of the problem.

[I] I am talking about c, ε_0, e, h.
[II] "The particle physicist is justified in ignoring gravity... and happy to!" (Ryder, Quantum Field Theory, p. 2).

And of course, claiming that the so-called Higgs mechanism or the Higgs particle would add anything relevant to the riddle of mass is bullshit. If you say something about the mass of an elementary particle, please predict a *number*! I don't give a damn about a "mechanism" that explains but does not calculate anything you can test. But the entire theorizing has shifted to such a vague and metaphorical way of reasoning, a takeover of cartoonish arguments that dominate physics today.

> *But when you cannot express it in numbers, your knowledge is of a meagre and unsatisfactory kind; it may be the beginning of knowledge, but you have scarcely, in your thoughts, advanced to the stage of science ... – Lord Kelvin, nineteenth-century British physicist*

Windbagging with Numbers

The ever-present excuse that things are complicated is best illustrated by huge (and expensive) computer simulations, which once in a while try to create the illusion of progress. A paper in *Science* a couple of years ago[4] claimed to have calculated the masses of neutrons and protons by means of "Lattice quantum chromodynamics" (a theory that, as we will see later had never produced a genuine result). Sadly, the accuracy was a factor of ten less than required even to distinguish proton and neutron, thus the whole thing is nothing but hot air. Despite the increase in computer power, nobody has entered the realm of testability in the meantime. One could not imagine better evidence for the vacuousness of that result.

Here we have, as the next and separate fault of the standard model, the incapacity of calculating numbers like the mass ratio of protons and electrons, which is 1836.15... Needless to say, calculating these values was considered a question of outstanding importance for example by Dirac, who spent years of his older life pondering about why Nature had invented this huge difference between the two most elementary particles, the electron and the proton. The modern picture that considers the proton to be a complicated mess composed of quarks and gluons flying about would have disgusted him.[1] The model of the proton has other serious drawbacks, one of them is the so called spin crisis.[5] Spin we may loosely imagine as the angular momentum of a rotating elementary particle which always takes the simple value $h/2$. The proton is probably a very simple thing, but in the context of the standard model, it is utterly incomprehensible how the proton – consisting of a bunch of particles each contributing their spin – can itself carry the rigid restriction of having spin $h/2$. This is just one of the dirty secrets (if it were a secret – but the repression is just psychological) that would render the model unacceptable for any reasonable person. Well, we are in the realm of particle physics.

Tweaking Towards the Metaphorical

> *The most striking feature of high energy physics is that it proceeded through a process of modeling or analogy[6] – Andrew Pickering, science historian*

Nevertheless there is a big riddle underlying the very notion of spin. The spin of elementary particles manifests itself in the fact that you have to twist them by 720 degrees instead of 360 in order to get them back into the original position. Rotations in three-dimensional space (which is what we live in actually!), which are denoted as group SO(3) in mathematical jargon, have precisely that intriguing property of needing an additional twin twist to become a mathematically simple structure called "double cover" SU(2). This is known mathematically, but not understood physically. We know how to describe it, but there is no clear explanation *why* the hell elementary particles must have spin.[II] If you believe in the standard model, spin is just an ornament, if not a nuisance that complicates the calculations. If the guys who call themselves physicists today had the same attitude as Ein-

---

[I] Another remark in the running for the silliest statement I ever heard is that it's no wonder one cannot calculate the proton mass, because there are "different mechanisms" at work. The mechanisms are stupidity and parroting.
[II] It does not follow either, as is often heard, from the Dirac equation.

stein, they would worry why Nature had invented spin. Nobody does. Instead, mathematical physicists, delighted by their *description* of spin, started to play stupid games with it.

When talking about elementary particles before, I deliberately named proton and electron as the very first, because they are the only *stable* ones – not an unimportant feature one might think. (Though today people rack their brains about lifetimes such as 10^{-25} seconds – we'll come back to that absurdity later.) The next important particle, so to speak, the neutron, decays after about 10 minutes (nobody knows why 10 and not 20) to a proton and an electron (and antineutrino). Both protons and neutrons are found in the atomic nucleus, and there the transformation can also go the other way, the proton turning into a neutron. It's not that this process is understood very well, but remembering the spin description, physicists decided by pure analogy to see the proton and neutron as the same particle being in two different states of "isospin." Not only do we learn nothing from this empty word about the decay (which is tantalizing to date), but it misleads physics to a merely metaphorical level, because isospin has literally nothing to do with spin. Having the mathematical toy at hand, in the 1960s, theorists[1] started to classify their accelerator products with "isospin" and other, similar artificial concepts, which later led to the ideology of "gauge groups" and abstract symmetries. This is one of the major sicknesses of particle physics: instead of worrying about the intriguing properties of truly fundamental (and observable) concepts like space, time and mass, a plethora of fancies, such as "isospin," "strangeness," "bottomness," "hypercharge" and similar baloney come to dominate the paradigm. Think about it for moment from the viewpoint of natural philosophy. How can one feel comfortable with a potentially infinite assortment of properties for elementary particles? Obviously, their invention required an unthinking sort of physicist. Einstein and Bohr, who used to discuss for hours about fundamental topics in the 1920s, hadn't deigned them a word of conversation.

---

[1] Sometimes it is falsely attributed to Werner Heisenberg, who in 1932 discussed the newly discovered neutron and its role in the nucleus.

The Higgs Fake

> *It is easier to recognize a prejudice in its naïve primitive form than the sophisticated dogma to which it often transforms.*[7] – *Erwin Schrödinger*

The Growth of Nonsense

There is a Chinese proverb that when rocks float on water, leaves will sink. Analogously, with all the real problems of physics suppressed, it is no wonder that a series of irrelevant "unsolved questions" should dominate the babble of today's science spin doctors. The so-called riddle of "confinement," why the three quarks in a proton can never be separated, while indeed unexplainable in conventional terms, is just a self-imposed problem indicating that the quark model is baloney.

You may also ask yourself why the cat has two holes in its fur in the same place as its eyes.[1] A similar pseudo-problem, which falsely suggests one is thinking, is, "Why there are three families of particles?" . The three "families" are just something to be thrown overboard once the next round of collider experiments need their results to be accommodated in a fourth or fifth family (neutrino counters have already entered that stage).

And eventually, the notion of four different interactions, with all its failed attempts at unification, is just a manifestation of a lack of true understanding of nuclear physics. (Heisenberg was the last one to state this. He died in 1975.) In 1930, people tried to understand the two interactions electrodynamics and gravity (then, the only ones). Postwar physicists, while deliberately ignoring gravity, invented two more interactions, as if this wasn't a sign of methodological degeneration.

What Really Matters

Let's consider the true riddles, the nature of time being a big one. The current notion of time as an invisible river that flows without relation to the surrounding matter is probably wrong. The Viennese

---

[1] A witticism of the German physicist Georg Christoph Lichtenberg (1742–1799) – very modern.

physicist and philosopher Ernst Mach (1838–1916) and more recently, the British physicist Julian Barbour, have made important points I cannot mention in depth in this book. Reversing the direction of time (as suggested by Richard Feynman) could make an electron turn into a positron, its mirror-image antiparticle with opposite charge. Antimatter, which was first discovered in 1932, is a big riddle because astronomers observe only matter and almost no antimatter in the universe. Common wisdom has a problem here, because it assumes (without any cogent evidence) that the big bang started with equal amounts of matter and antimatter.

Theorists soon came up with fix: Due to an "asymmetry," the early universe by chance happened to be in an unbalanced state that made the universe turn into the thing we observe today. The science historian David Lindley commented wryly: "Asymmetry was introduced in the theory and carefully adjusted for no other reason than to produce the desired answer." Now savor the groundbreaking enlightenment of theorists as to what caused the symmetry to become asymmetric: a "symmetry breaking." It is precisely such empty claptrap that had physics turned into the unscientific state of semiotic subtleties.

The anger induced by the modern nonsense occasionally distracts me, but let's get back to the real problems: it is much more likely that the tantalizing rareness of antimatter is just related to the direction of time. If you ponder more intensely, you will realize that we don't have a clue yet about the true nature of time. A natural measure of time is the decay rates (half-life) of all unstable particles, they are natural clocks. But can we calculate these half-lives, and if not, why not? And an even more irritating question: Why did Nature provide all these unstable particles after all, instead of being happy with an unchanging assortment of particles? There must be an answer, we just don't know it. Einstein used to express such questions with the phrase: "I want to know whether God had a choice." But the very reason for radioactivity, despite it being an intensely studied research field for a century, is utterly unknown. Our current understanding suggests that it is just a complicated quirk of Nature, but this suggests to me our understanding is sort of quirky.

Come out with a Number

Today's scientists got widely used to cheap patches when it comes to fixing some contradiction in an ad-hoc manner, but the real problems fall into oblivion. Take, for instance, the fine structure constant, a combination of the constants c, e, ε_0 and h. The number 137.035999... is, according to Richard Feynman, "one of the great damn mysteries of physics" and he recommended all good theoretical physicists should "put this number up on their wall and worry about it."

Since Feynman wrote this 30 years ago, none of the 10,000 particle physicists has made any damn progress in that question. It might be difficult, yes, but those who didn't even spare a thought about it in the past 30 years (I bet this is the vast majority of CERN theorists) please don't tell me that you deal with something fundamental. Physics is about the big questions, not about fiddling around with 150 parameters.

A big issue, admittedly seen as such by all physicists, is the incompatibility of quantum mechanics and gravity (apart from string theorists' ridiculous boasting of having explained the "existence of gravity." – they need to come back to earth and fall on their butts to get a proper understanding of gravity). The problem is mostly (ab)used as a mathematical playground, and often garnished with the drivel "hierarchy problem." The problem is just this: The ratio of the electric and gravitational force of a proton and an electron that form a hydrogen atom is a huge number – 10^{39} – and nobody knows where it comes from. Period. Beware wannabe unifiers: either you explain that number or you had better shut up. An intriguing idea with respect to this riddle came from Paul Dirac in 1938. He observed that the radius of the universe, divided by the radius of the proton, yields roughly the same number. Strange enough, but there is more to come. The square of that large number is approximately the number of protons (10^{78}) in the universe, a tantalizing coincidence. If there were a hidden law, it would not only screw up the cosmologists' current picture of cosmic evolution, but in the first place it would disturb the particle physicists' firm belief that the mass of elementary particles cannot be explained (instead of considering the possibility that they are too dumb to understand the issue). No wonder that Dirac's ingenious observation, called the Large Number Hypothesis, is usually dismissed as "playing with numbers," "numerology," or "esoteric." Yet it is, among all the crap that has been written for seven decades about unification, the only quantitative idea about this biggest riddle of physics.

As we have seen when pondering simplicity, physics must focus on the role of fundamental constants when dealing with fundamental questions. Related to Dirac's thoughts, the American astrophysicist Robert Dicke in 1957 developed an extremely interesting concept which relates to an idea of Einstein, who in 1911 dealt with a variable speed of light.[8]

Exceptional Ignorance

Today's physicists are not only ignorant about these ideas, but they are actively distributing their mindcuffs, like three theorists who published a wacky ideology, "only dimensionless constants can have a fundamental meaning," which summarized their discussion in the CERN cafeteria.[9] Oh, had Einstein had the opportunity to listen to their half-assed thoughts, while having a cappuccino there! Surely, he would suddenly have understood how misguided his 1911 attempts on a variable speed of light were. And if Maxwell also had his coffee close to the accelerator, thanks to the three avant-gardists he would never have bothered the world with his theory of electromagnetic waves – born out of pondering on the "dimensionful" constants ε_0, μ_0 and c.[1]

To summarize, the standard model of particle physics has nothing to say about the contradictions of electrodynamics, nothing about masses, nothing about ratios of masses, nothing about lifetimes, nothing about the fine structure constant, nothing of the relation to gravity, nothing about the deeper reason of spin, nothing about radioactivity, nothing on the nature of space, time and inertia. Whatever

---

[1] Historically denoted in a different manner.

The Higgs Fake

pseudo-problems particle physicists may tackle, a theory that says nothing about those fundamental issues is crap.

Alexander Unzicker

CHAPTER 3

HISTORICAL IGNORANCE

WHY PARTICLE PHYSICS CONTINUES TO REPEAT THE OLD ERRORS

Insanity in individuals is something rare – but in groups, parties, nations and epochs, it is the rule. – Friedrich Nietzsche, German philosopher

Besides the issues of simplicity and fundamental questions, today's particle physicists show a total lack of understanding of how science works and has always worked. History is not just a dull thing of which you can pull off your introductory anecdote of a plenary talk. To see the mechanisms that have been repeated many times greatly helps to evaluate the current situation. It would help. In high energy physics particularly, though, it is also the overestimation of the news of the day, the unhealthy focus on the present, and a good deal of arrogance, that contributes to the absurd complication of the standard model that disregards the big questions.

And whenever scientists (if they reflect about the development of their field at all) try to give an historic account, they prefer to rush through a short version of it, just out of the necessity to date discoveries. This is usually a rose-tinted view. History in general suffers from the attitude of retrospectively inventing narratives, just to put the sequence of events in a logical chain. That is a delusion. Only detailed accounts reveal how many options were available that were forgotten later and how many times history could take just another path. Unfortunately, there are very few good accounts of the history of particle physics (like Andrew Pickering's), and, needless to say, these are usually unknown to today's busy researchers.

Beware of Broader Education

Considering science in general and on longer time scales, there is much that particle physicists could learn from history. In particular, the philosopher Thomas Kuhn, in his epoch-making treatise, *The Structure of Scientific Revolutions*, pointed out the mechanisms that occur whenever humans do science. Most importantly, increase in knowledge is never linear in time. Periods in which progress seems to be incremental are interrupted by scientific revolutions where a good deal of the previously existing knowledge is destroyed. This asymmetry in time appears to be a universal mechanism; in stockmarket rates, for example, there is mostly incremental growth and sometimes crash-like decrease, rarely the opposite. Thus the more or less incremental construction of the standard model in post-war physics suggests that it is not healthy for purely historical reasons. No wonder that Steven Weinberg, one of the model builders, was so uneasy with this painful prospective that he felt he had to attack Kuhn in a poorly-reasoned article, "The Revolutions That Didn't Happen."[10] Well, they still have to. Nassim Taleb calls such an attitude the *turkey illusion*: Until the day before Thanksgiving, all the evidence says that everything is ok.

It Has All Happened Before

The prototypical example of an unhealthy theory was Ptolemy's model of astronomy with the Earth at the center, followed by the Copernican revolution that trashed it. Today we ridicule the idea of the Earth being the center of the universe, but at the time it seemed perfectly reasonable. Didn't everything fall down on Earth? See?

Evidence for the opposite, like the tiny shifts of star positions, were not observable at the time. Eventually, the retrograde motion of other planets, such as when we pass by Mars on the "inside lane" around the sun, had a satisfactory explanation: people imagined that on top of a circle surrounding Earth, there was another circle mounted which defined Mars' orbit, a so-called epicycle. It sounds silly today, but the leading scientists of the time were convinced of it. Just a slight, unmotivated complication could cast some doubts on the picture. Remind you of something? Since the epicycles didn't fit perfectly, the center of the main circle was shifted a little (the so-called excenter), to account for the difference. And when, as if it were a side effect of the previous cure, the velocity turned out to be wrong, another arbitrary point, the equant, was postulated from which the motion appeared uniform. Since then, the epicycle theory has become a synonym of thoughtless complication. But there *was* "evidence" that supported it, even fairly precise evidence. (Again, does this remind you of something?) Particle physicists never think of the precision of medieval astronomy when they repeat their gospel of the "excellent agreement with the data" of the standard model. If you have a bunch of parameters to fiddle around with, it is easy to get there.

Behind the epicycle model there was also an unquestioned belief, such as circles being the most natural, symmetric things the laws of Nature had to obey. That "gauge symmetries" in particle physics have taken a similar role (while being blatantly unjustified by direct observation), should leap to the eye. And we will later look at the sequence of mismatches with the theory, which led high energy physicists to postulate one new particle after another. The epicycle model, thoroughly inspected, should make every particle physicist notice he or she is trapped in the same mess. As Julian Barbour points out in his excellent book *The Discovery of Dynamics*,[11] back then physicists also fooled themselves by seemingly surprising, though coincidental, "tests" which had a remarkable "precision." But no particle physicist would study that closely. They consider the epicycle model an isolated event, rather than just one example of a mechanism from which something could be grasped. When you remind them of history, they often come up with the dumb statement, "It's much harder do make progress today..." It is like saying "unfortunately, the roads have vanished" when you are lost in the jungle. Physics has lost the scent.

> *The agreement of a dumb theory with reality says nothing.* – Lev Landau, Russian Nobel laureate

Perverting Kuhn's Insight

If particle physicists talk about a future paradigmatic change in a Kuhnian sense, they misapply it in a grotesque way by stating that particle physics may undergo a transition, such as from Newtonian mechanics to Einstein's relativity. This is nonsense. Einstein's general relativity refined Newton's law of gravitation,[12] but it did not simplify it in the sense that it needed less parameters. Newton's theory never underwent the piling up of absurd complications that we know from the standard model.[1] Nevertheless they dare to compare their illogical turmoil to Newton's clear thoughts, hoping that the standard model will be "embedded" by a future theory of the sought-after new Einstein. Wishful thinking. It is rather a Copernicus or a Kepler that is needed. All that will remain after the crash of the standard model, when the thin fouling is brushed off the rocks, is quantum mechanics as developed in the 1920s. But this is a much too scary perspective for particle physicists to let it even faintly cross their minds.

Besides the epicycle model that dominated astronomy for fifteen centuries, history has instructive examples on a much shorter time scale. Just look at the story of Otto Hahn and Fritz Strassmann in Berlin in the late 1930s, which led to the development of the atomic bomb. Though they had already split the atomic nucleus with their neutron experiments, it took them almost five years to realize it. To explain the obviously complicated experimental results, Hahn assumed that new heavy elements, called "transuranians," had been formed when the neutrons (which had induced the nuclear fission) approached the uranium nucleus. To account for the unexpected

---

[1] There was just one observational mismatch, the anomaly of Mercury's precession.

variety of substances, Hahn introduced a series of secondary ad-hoc hypotheses, such as the same isotopes having different lifetimes. None of those assumptions was totally unrealistic, but at the same time they were unsupported, while a series of accompanying puzzles was ignored. I don't understand why it does not dawn on particle theorists that their bunch of quarks are inventions analogous to the heavy elements. They have long since lost an overview on the unmotivated complications in their theoretical description.

Consider however that the solution for Otto Hahn – nuclear fission – was such an obvious, simple and easily testable case, yet it nevertheless took until 1938 until he eventually saw it. It is obvious that today we overlook something like nuclear fission or the elliptical form of planetary orbits (rather than them being circles). And of course, the underlying problem in particle physics, which probably goes back to the 1920s, will be not be such a simple one. Accordingly, we must be prepared for the dead end we are in to be a long one. And the errors may be rewarded. In 1938, Enrico Fermi got his (experimentally well deserved) Nobel Prize for the mistaken theory of transuranic elements...

No Way to Coexist

The fact that an opinion has been widely held is no evidence that it is not utterly absurd; indeed, in view of the silliness of the majority of mankind, a widespread belief is more likely to be foolish than sensible.[13] – *Bertrand Russell*

The good news is that you can have a nice career in such a dead end. Once we realize that post-war particle physics is in a Kuhnian crisis, it follows that already a dozen Nobel Prizes have been handed out that did not advance physics in the slightest; to Alvarez, Gell-Mann, Zweig, Richter, Ting, Glashow, Salam, Weinberg, Rubbia, Lederman, Schwarz, Steinberger, Friedmann, Kendall, Taylor, Perl, 't Hooft, Veltman, Gross, Politzer, Wilczek, Nambu, Kobayashi, and Maskawa. I know it sounds pretentious, but consider that the reasoning of Newton, Maxwell, Planck, Einstein, Bohr, de Broglie, Dirac,

Heisenberg and Schrödinger[1] is just incompatible with the contemporary stuff. There is no way to carve both groups into a Mount Rushmore of physics. And if one weighs the results instead of just counting them, it is obvious which ones we have to dump. "The old days are gone" some may retort, in view of the outstanding discoveries of the past, but in the first place, *reason* has gone today.

It is an unpleasant scenario for the researchers involved, but the mechanism described by Kuhn, that sequence of small reasonable steps that slip into unreliable complexity followed by violent crises, is present everywhere and on all time scales. Consider the case of the phlogiston theory in chemistry that hindered the discovery of oxygen, or even more drastically, geologists' denial of continental drift for half a century. Alfred Wegener, a meteorologist (and therefore not taken seriously by the experts) who in 1912 had drawn the right conclusions from the African and South American coastlines and accompanying fossil evidence, was ridiculed and the puzzling evidence was explained away by a series of complicated (retrospectively awkward) secondary hypotheses. Wegener had already died when in the 1960s it dawned on the community of geologists that they had been mistaken – and it was not due to their own flash of genius but due to the inescapable evidence of seafloor spreading.

The Graveyard Solution

The problem is that there isn't even a possibility to prove particle physicists wrong today. The longer a dead end, the harder it is to get out, but the most critical point, actually a point of no return, is where an entire generation of researchers has lost its way – those who knew about the line of retreat are dead, and voices from outside the community, like Alfred Wegener, fall on deaf ears. It is disastrous when the twerps, with the authority they inevitably gain as senior scientists, decide how the next generation is to be educated. Whenever the rhythm of discoveries becomes slower than homo sapiens' life span, there is a problem. Max Planck once said that, in these cases, science

---

[1] Omitting a dozen of the Nobel Prizes awarded to experimenters in the early twentieth century.

takes a very slow pace: from one funeral to the next. If the model is institutionalized as it is today, that doesn't help either.

Generations of physicists have grown up now who had to rehash the standard model in their exams. And there is no reason to suspect that the most intelligent ones chose particle physics as their field of research. The geniuses are long dead. Whoever in the 1970s or '80s (when the model already showed overwhelming complication) went into high energy physics, knew what was waiting for him. People who even faintly respected Einstein, Dirac or Schrödinger's views on the laws of Nature were not among them. On the other hand, due to its proximity to the powerful nuclear physics, the field was too well funded to die out. Thus it is kept alive by huge transfusions of cash that hide the intellectual degeneration, while even developments that are declared as progress happen incredibly slowly. (If anything, particle physicists should realize this, even when calling their non-results results; revolutions happen quickly!)

The history of science is not separate from the rest of history, however. While the development of particle physics was largely incremental in a Kuhnian sense in postwar physics, there was a historical disruption. When the Nazis destroyed the intellectual centers of physics, Einstein's emigration was a symbol of the end of the tradition of physics rooted in philosophy in Europe. And there was hardly an event that had more impact on modern physics than the dropping of the atomic bombs in August 1945. Science had become powerful. The accelerators that had unleashed the nuclear power and shifted the power on the globe developed their own life. Earnest Lawrence's palm-sized cyclotron has grown to the 27-kilometer circumference of the ring at CERN. But it is not technology that provides insight. The impressive apparatuses, while holding a persuasive fascination of their own, helped to gather people. But all this cannot hide the fact that physics turned into another science in that period – if it is still a science at all.

Physics has undergone a transition since 1930. Unexpected surprises in small experiments became a forced search using giant machines, the mavericks of the past have been replaced by big science collaborations: pursuing fundamental questions has turned into a fashion of pragmatic administration of results. Physics, in short, has turned into a high-tech sport,[1] that for historical reasons has stayed the name. Some claim that "all the cheap experiments have been done" when advocating the next round of a pointless higher-energy upgrade. But there is just no room for creative intelligence in this business. If there had been no high energy physics in the last 60 years, the planet would probably be better off.

Black Swans and Just Black Ducks

Almost all of the greatest discoveries, those which significantly changed the path of history, came unexpected, or, as Nassim Taleb phrased it, as black swans. The message of the virtue of the unexpected has somehow sunk in within the research community, but in a very weird way. Many see *any* surprising outcome of an experiment as something positive. Sorry, but why is something you don't understand always positive? Any failure to predict can be declared a success, besides the "usual" success of being vindicated – a good strategy. The really unexpected things you cannot actively generate. When you build a huge collider with particle energies that surely go beyond your understanding, it is almost guaranteed that you will discover *something*. But this is where discovering something unexpected is perfectly expected.

The history of high energy physics of the last few decades shows that there is a gullible attitude to planning the non plannable, and some find particular pleasure in this kind of coquetry. There is hardly a figure that better represents the thoughtlessness in high energy physics than the British theorist John Ellis. His career consisted of distributing the latest fashions, be it neutral currents, new quarks, or supersymmetry. Never having found out anything, he serves now as a caricature of a theorist, a kind of mascot for the media. In 1982, CERN had informally communicated its results to the British Prime Minister before the official release. (What science, if not corrupt, needs to stay so close to politics?) Ellis off-handedly told Margret

---

[1] The detectors make use of physics, but in the same manner as the car or airplane industry does.

Thatcher that CERN, in its future experiments, hoped to discover the unexpected. It's a good thing that Thatcher's era is over, but she did have what you may call common sense, and she retorted, "Wouldn't it be better if you found what you expected?"
With respect to the standard model, particle physicists have developed this ludicrous attitude we are all used to: either we find a wonderful confirmation of our models or – even more exciting – a contradiction. In any case a tremendous success. This is the prototype of a nonfalsifiable model, according to Karl Popper – the opposite of science. With the same kind of wacky philosophy, theorists had insured themselves until right up to the "discovery" of the Higgs boson in 2012. The standard model, with its complication, its ad-hoc patches of contradictory data, and its lack of predictive value, is the best example of a Kuhnian crisis, but at the same time the people involved have the least prospect of getting out of it. It is high time that the standard model became history.

CHAPTER 4

EVIDENCE SHMEVIDENCE

ENDLESSLY REFINED SIEVING: HOW PARTICLE PHYSICISTS FOOL THEMSELVES

> *The establishment defends itself by complicating everything to the point of incomprehensibility.* – Fred Hoyle, British astronomer

The arguments regarding simplicity, fundamental problems, and history mentioned in the previous chapters should be convincing, nevertheless they may be considered somewhat general (in particular to the specialists prone to suppression). Thus in this chapter we come to the very mechanism with which particle physicists fool themselves.

The history of particle physics is a characteristic sequence of events. The pattern consists of (1) developing theoretical fashions too vague to be scrutinized, (2) letting particles collide in accelerators, (3) declaring every outcome that is poorly understood as the discovery of a fashionable new particle, (4) filtering that particle in the next round of experiments as "background" (which is then uninteresting), and (5) doing another experiment based on the very same idea, but at higher energy. Increase the energy until you don't understand something, and so on. It is a rat race. We may compare it to investigating a gravel bed with all sorts and sizes of gravel. The biggest stones are easy to detect and are effectively filtered by a coarse sieve, which corresponds to a small, cheap accelerator. What falls through, the unexplained, is labeled according to the theoretical fashions of the time. Since it needs slightly more sophisticated detection methods, a finer sieve in the form of a new accelerator is built. Once detected, however, physicists lose interest in the new particle which is considered "background" in the following. The finer the sieve, the more energetic the collider, and the more stuff you can crash (called luminosity) the more data you can gather. Whatever falls through the sieve, every outcome, however contradictory it might be, is a new result to celebrate. And then they need a finer sieve, and so on. But particle physicists have long since reached the stage where they just churn the sludge, trapped in a morass of accelerator data that has become impenetrable by clean methods. The background has become overwhelming with respect to the desired results, outshining any reasonable signal. Particle physicists are like astronomers working in daylight.

To extract signals, technology has to be pushed to the limits, but this is anything but a scientific revolution. Still, they call that neverending futile loop scientific progress. The absurdity is best illustrated with the phrase "yesterday's Nobel prize is today's background" espoused to me by a particle physicist from CERN in a panel discussion (yes, that was an eye-opener). Where is this going to finish? They think they can continue with their nonsensical dance forever (and be fed by the Nobel foundation). How can one, thinking on the long term, make use of such an idiotic strategy? But the last reflective exemplars of that species of high energy physicists have long died out.

Tunneling Views

In the last decades, a cascade of increasingly rare effects that needed colliders with ever more power were forcibly interpreted as new particles: the bottom quark, W and Z bosons, the top quark, the Higgs boson. The sieve is getting finer and finer, and it is ever more likely that artifacts of the analysis come into play that do not genuinely exist in Nature. In our metaphor, the finer and finer grains, rather than existing beforehand, would be produced by abrasion during the process of sieving, when the nuggets have long gone.

To focus on higher energies, physics has developed an array of techniques that, however, risk masking potentially interesting effects: While the fixed-targets experiments in the early days were more interesting in many aspects, since the 1970s the dominating philosophy demands two colliding beams of particles – the energies are higher in this case. Correspondingly, physicists are interested in "hard scattering" events, particle debris that comes out from the collision point almost at right angles, with a high "transverse momentum." There is no fundamental reason why this should be more interesting than all the other reactions, it is just a consequence of the above futile philosophy of looking for the more exotic. Some 99.9998 percent of the data are discarded right after the collision because they are considered uninteresting in terms of the model assumptions (in the meantime the physicists justify it with the amount of data they can't store otherwise). No wonder that this method of "triggering" was invented during the search for the W boson in the early 1980s, in the methodological dark ages, as we shall see later. In essence, triggering is just utter disrespect for the true outcome of an experiment. High energy physicists think that all that is relevant has to comply with their prejudices about Nature, and they continue to narrow their view until they go blindfolded. It is guaranteed that when performing such a strategy no truly surprising result can come out. It has been dumped at the outset. Modern colliders are the most mind-cuffed experiments ever conducted.

I don't understand why the strategy of hunting ever more rare effects, while discarding everything that has been found before, is not recognized to be as deficient as it is. It's a never-ending process necessarily leading to nothing but complication. Adherents of homeopathy claim that the greater the dilution, the more effective the drug, particle physics say the rarer the event, the more important it is. It is an ideology, nothing else than a steady transition to nontestability: while scrutinizing this almost-nothing they head for knowing everything on nothing.

The Haywire Haystack

The identification of the allegedly new particles has become a task so complex that it is methodologically hilarious. It's not that you have to find a needle in a haystack, the needle is in a needle stack, as CERN director Rolf-Dieter Heuer phrased it, apparently being proud of having developed a metaphor for absurdity. Joe Incandela (speaker of the CMS experiment), when talking about the Higgs discovery, referred more technically to a photon background of 10^{12}, and said it was like identifying a grain of sand in a swimming pool full of sand. No streak of doubt in the data analysts' abilities seemed to have bothered him either. *Physicists see Mississippi water level rise because a boy peed in it.* Believe that headline? If it comes from a scientific press conference, yes, we do. But let us address more seriously this ludicrousness. Consider the half-life of W bosons or top quarks (both of which enter the Higgs analysis) of some 10^{-25} seconds. It means that the poor particle, during its lifetime, cannot travel more than the radius of a proton. It will never get out of the mess of the collision point where 600 million protons crash into each other every second, never get into any detector (!), never cause any physical process that can be clearly identified unless you pile dozens of theoretical assumptions on top of each other. But high energy physicists claim to know precisely what's going on.

> *Most particle physicists think they are doing science when they're really just cleaning up the mess after the party.* – Per Bak, Danish theoretical physicist

Turning Physics into a Casino

When physicists try to separate 100 billion pairs of photons of "background" from the one pair which originated from the decay of a Higgs boson, it is rather contrived to speak of an "observation." But if you want to continue where it is senseless to continue, use computers. Of course, I do not object here to the use of computers in general for analyzing data. But if you have reached the experimental limits, there is no clean way to work around it by using virtual reality. Not coincidentally, such "Monte Carlo" simulations were first used in the Manhattan Project when physicists built the atomic bomb – because in that situation, no previous "experiment" was indeed possible. Monte Carlo simulations can be useful when you want to make something work which is too complex to understand, but not if you want to understand something that is too complex to work with. "Event generators", as particle physicists call them, are used where the usual theoretical calculations (which are already superficial descriptions) don't work anymore. They constrain themselves to simulate the collision (with some poorly justified simplifying assumptions), at a still inferior level of understanding. Such gullible modeling is again reminiscent of Nassim Taleb's description of economic models based on Gaussian statistics that never work.[1] The problem is that physicists not only have no skin in their taxpayer-funded game, but due to the missing link to anything real there are no stockmarket crashes that could illustrate the usefulness of what they are doing.

Cherry-picking Reality

Naively, one might think that the detectors register all the particle debris splashing out of the collision point and analyze it. But it's not quite like that. Instead, there are multiple layers of theoretical assumptions sandwiched between the event and the final outcome, each of which is realized by a computer code that calculates how many of this-and-that type of particles are produced, how these hypothetical particles fly about and behave during their mostly minuscule lifespan, and how all the products transform, sometimes after several cascades, to the signals that should eventually be measured in the detector. And if you have dozens of particles with hundreds of assumptions about their properties worked out in millions of lines of

---

[1] Monte Carlo simulations were used for simulating the performance of financial derivatives. There is reason to suspect they are obnoxious for physics as well.

impenetrable computer code, then there is plenty of stuff you can fiddle around with to make the outcome agree.

Look at the disproportion: the detector signal, ultimately, is something very simple, not to say banal: energy, charge, and the track of a particle. The hypothesized scenery in between is tremendous: muons, taus, all sorts of neutrinos, top, bottom, charm and strange quark pairs, W and Z bosons… all of them are, essentially, freely adjustable tuning screws that can be set to describe the result in terms of what you want to see. This methodological pitfall is particularly pronounced when neutral particles different from the photon are in the game. They don't interact with the electromagnetic calorimeters (energy measuring), the only sign of their presence is the sudden appearance of a bunch of particles (a "hadronic shower") at the moment of their decay. But in principle (experts will deny this with nitpicking arguments) *every* neutral particle can do that.

It is essential to understand (though high energy physicists never will) that at this point there is no objective interpretation, no unambiguous reality of an "observed" experiment but an inevitable invasion of theoretical assumptions – in this case, of the grubby standard model. We will come back to this point in more detail, but please keep in mind that particle physics has long distanced itself from the ideal of a theory clearly distinct from the experiment that acts as an impartial arbiter to judge theories.[1] People today muddle along and don't even know any more what is model and what is fact.

Inviting to Fancy

A particularly bizarre method of particle identification relies on "missing transverse momentum." Momentum conservation (an elementary law), loosely speaking, states that the debris after a collision is supposed to fly off in all directions, much as a snowball slammed into a wall will not just splash upwards or to the right. Such an asymmetry is occasionally observed in the collisions, and particle physicists make a big fuss about it, concluding that a rare particle,

---

[1] Andrew Pickering calls this "naive realism.!

such as a neutrino, went off undetected and kept the momentum balance. If the situation does not allow a neutrino going off, naturally, the door is open for further interpretation. An essential element of detection of the W boson relies on such missing transverse momentum. However, such an interpretation (after one has observed some elementary physics constraints such as energy momentum conservation in the "center of mass" frame) opens the floodgates for any kind of fanciful speculation, while in the extensively complicated collision there are plenty of reasons why the momentum could be missing: detector malfunction, interaction with the rest of the particles in the beam, with the beam tube, secondary interactions, lack of theoretical understanding of the multitude of processes involved or simply nasty little details one hasn't thought of before. All this is "negligible" according to conventional wisdom, but who knows? Once you interpret the deficit of transverse momentum, the methodological problem is that you actually get evidence from something which is missing, that is, seeing by non-seeing. Unfortunately, the subtle but infinite potential to fool oneself is one of the grave absurdities physicists (and even philosophers) never reflect on. And of course, it is impossible to separate theoretical assumption from (non-) observation. The standard model of particle physics has long since been in a state where its own alleged results have entered by the back door as assumptions, watering down the experimental method to the muddy juggling that is practiced today.

One thing that greatly facilitates the flourishing of murky concepts in particle physics is lack of accuracy. Experiments in atomic physics measure the respective constants to an accuracy of ten decimal places: Einstein's general theory of relativity is tested to the fifth decimal place. The energy revolution of the LHC instead is at best 6 percent. High energy physicists will now bemoan that it is the best possible (I don't deny they did their best), but it is still, in absolute terms, poor. It is true that electron-positron colliders have a better resolution, but the "precise tests" of the standard model everybody boasts about are merely propaganda. And there is a lot of cherrypicking. If a result contradicts the model, it is euphemized as "known tension,"[14] e.g., "in the bottom sector." It is like saying, "The patient is in perfect health, just some organs are damaged." And should the "tension" become too great, there is always a suitable *extension* of the model at hand.

In the Box

Since we hear so often that the standard model is "precisely tested" (what a bizarre statement in its universality, given that one cannot even define what the "standard model" is), let´s see what remains when we leave aside the self-deceiving fiddling with free parameters. Here is the best test, if we are being benevolent. The argument that involves W and Z bosons, requires a short deviation.

A true test means that you can compute a quantity from data of different fields, and this can be done for the so-called weak coupling constant g – according to the standard model, the constant of Nature responsible for the radioactive (beta) decay. g can be computed from the mass and half-life of the muon, an unstable particle best described as a "heavy electron." The muon decay can be measured quite accurately, and yields[1] a value of $g=0.6522$. A second way to determine g is using the masses of the W and Z bosons, for which a remarkable accuracy is claimed (80.387 and 91.2187 gigaelectron volts respectively). This yields g=0.6414..., agreeing to within 1.7 percent. This is not extraordinary (theorists will argue that it can be improved by further corrections involving another bunch of particles), but it doesn't sound too bad. It is worthwhile, however, to look at how the masses of the W and Z bosons are determined. Remember that a W boson signal was nothing but a fast muon (or electron) and a neutrino, of which the energy cannot be measured! What physicists do now is to estimate the missing energy (of which, additionally only the part transverse to the beam is known) and compare it with a "Monte Carlo" computer simulation of W bosons with different masses.[15] But how should such a rare decay, with so many unknowns entering the analysis, allow the detector's energy resolution to reduce from 6 percent to 0.02 percent? It seems quite an ambitious claim,

---

[1] Technically, g is obtained from $((8 \pi)^2 \, 48 \, h \, m_W^4 \, c^8 \, / ((m_\mu c^2)^5 \, m_\tau))^{1/4}$ in Bleck-Neuhaus, formula 12.6.

and it is untenable that such computer analyses cannot be scrutinized by the public. All this (and there are many such opaque procedures) has to be put on the internet in detail to allow for future independent repetition—that's how science is supposed to work.

Again, let us assume all systematic errors are under control and the claimed accuracy of the W boson mass *is* correct. In the end, the two different determinations of g are not as independent as they seem. The first measurement involved the mass and half-life of a muon, and the latter procedure again dealt with the mass of the muon and its alleged companion the neutrino produced some 10^{-25} seconds after the birth of the W boson. Thus in both cases we measure properties of the muon! It is possible that the theoretical formulas yielding the approximate coincidence just mean that there is a relation we do not understand, but we get tickled by it like a dog chasing its tail. This paragon test of "electroweak unification" actually deals with the unification of the muon with the muon.

And there is another pitfall of this test on a more fundamental level. We are talking about the "weak" force allegedly responsible for the decay of a neutron. The muon has a very indirect relation to the weak force, since during its decay it produces two neutrinos, which in a very rare process may generate the beta decay. But why does the model deal with this lower-rung muon instead of predicting the decay time of the neutron, a much less elusive constituent of matter? The standard model is cherry-picking its tests among the rare events, while it beats around the bush when it comes to explain the important things. It is a highly selective reality where these guys have made themselves comfortable.

Nothing is so difficult as not deceiving oneself – Ludwig Wittgenstein

Not an Insinuation ... But no Security Either

It is also obvious that such a complicated data analysis is extremely vulnerable to errors – and fraud. I am not insinuating anything, but it is the sheer number of people contributing that makes the common effort sensitive to all sorts of manipulation. If anyone of the hundreds of programmers had bad intentions, he probably could fake a lot of results. In the financial sector opportunity made the thief, and in each of the numerous cases it happened we were told that nobody could have foreseen it. Indeed, who could? But the lack of control is obvious, and scientists are less susceptible to mistrust their colleagues than bankers (justifiably).

Still, I think it is much more likely that unwanted errors, computer bugs and instrumental artifacts may screw up the analysis. Around 1995, several groups – independently, as they believed – found a striking contradiction to the standard model in the form of a decay of the Z_0 particle into a bottom-antibottom pair, until David Browne, a researcher of the Lawrence Berkeley Laboratory, eventually discovered that a wrong simulation had been fed into a central database. And there are multiple examples of artifacts in history (some of them quite embarrassing) being interpreted as important discoveries: the 17 keV neutrino, a mirage particle confirmed by six independent research groups, the "positron lines", an anomaly that kept the *Gesellschaft für Schwerionenforschung* GSI (one of the most prestigious institutes in Germany) busy for several years until it was eventually identified as an artifact of triggering and unjustified data selection,[16] the comical drama on superluminal neutrinos from CERN produced by a defective plug, and so on. On July 4, 1983, exactly 29 years before the notorious Higgs seminar, a press conference was held at CERN to report the tentative discovery[17] of the then long-sought-after top quark. It turned out to be wrong. All this is forgotten only too willingly in the retrospect narrative – or, worse (if you remind the parties involved), presented as an example of how well the scientific method worked. Too often it was damn good luck, however, and it took an embarrassing time span until the error was identified as such.

The Gain of Being on the Right Track

Here we encounter one of the most underestimated distortions that bias science, the reaction to an unexpected result. The mechanism is lucidly outlined by Andrew Pickering[18] and goes as follows: However carefully the experimenters have worked, an unexpected result could still be due to an error. Now let a theorist enter the sce-

The Higgs Fake

ne, who declares that exactly this result confirms his theory. This creates new options, welcome to both: The unexpected is not your bug, but a feature of nature, to be explored further by the brave experimenter, while the theorist can justify further development of his pet theory with "experimental hints" and raise new questions experimenters can tackle. This is the essence of what Pickering insightfully called the "symbiosis" of theoretical and experimental practice, and therefore, was utterly ignored by particle physicists. Bury your head in the sand if someone tells you how you are fooling yourself. Mind in this context *how* scientists are allowed to be mistaken. Systematic errors can ruin your career, even if they inevitably occurred during a thoroughly conducted experiment. On the other hand, nobody frowns at a theorist's proposal, however wacky the assumptions it may be based on – it's a peccadillo. All this favors the "symbiosis."

Let's be honest: the experiment with its dry, technical details, that requires such anal capabilities as care, attention and diligence, is a boring thing compared with the glamorous, creative inspirations about the design of the universe the theorist likes to dedicate himself to. This is also the reason why there is so much theoretical shit in arXiv.org, rather than detailed but accessible studies of detector responses, calibration problems, and so on. But no one will ever get a Nobel Prize for discovering that photomultiplier tube noise unexpectedly increases at a certain low temperature.[1]

An honorable exception to this unpardonable disregard is Peter Galison's book, *How Experiments End*, a highly interesting description of high energy physics in the 1970s. When discussing the "weak neutral current" discovery, Galison also addresses the sociology of decision making in science. The sociological aspect of the above "symbiosis" as a kind of "insight" is unidirectional, a misfit is always seen as need for further elaboration of the theory.[19] Theorists and experimenters would never go back and agree that what they did for a couple of years was wrong.

---

[1] Just a little, but a real example. I always wonder how many such unexpected effects may be hidden in the equipment of the big particle detectors.

Alexander Unzicker

Letting the cat outta the bag is a whole lot easier than putting it back. — *Will Rogers*

Fruitful Symbiosis

It is a characteristic of this symbiosis of theorists and experimenters that they are interested in phenomena developed more by the respective practice than by nature. "Jets" for example is a jargon denoting two showers of particles streaming, as required by the conservation of momentum, in opposite directions. But it is a purely descriptive feature of particle collisions, nothing indicates that there is anything fundamental behind it. Yet you can wonder about "mono-jets" and "three-jets", which seemingly violate momentum conservation and therefore *may* point to new particles – or to poor statistics or to detector malfunction. Such things have created extensive fashions in the community, in particular when linked (in some allegorical way, because clear-cut predictions don't exist in the field) to a fancy feature of a fashionable theoretical model. All these concepts are products of social practice, as Pickering says, and therefore, seen on a fundamental level, bullshit. No wonder Einstein, Dirac or Schrödinger never wasted a thought about this industrial production of unintelligible stuff.

Let's come back to the possible reasons for science losing its way. Fraudulent action seems unlikely, artifacts or mismodeling fairly likely, though there are no laurels to gain if you try to demonstrate it. Still, everything could be ok with the data analysis (though with a much lower probability than CERN folks claim). It is the very system of continuously looking after the smaller and smaller effect and declaring it a discovery that is, at heart, flawed. It's not the problem that an artifact is occasionally declared as signal. The problem is that the entire system is built on such a sick paradigm. Until particle physicists don't understand that their philosophy "yesterday's Nobel is today's background" is deeply foolish, they will lead physics just deeper into the mess.

CHAPTER 5

THE BIG PARROTING

HOW INCESTUOUS EXPERT OPINIONS SPREAD

> *Few are able to calmly pronounce opinions that dissent from the prejudices of their environment; most are even incapable ever to reach such opinions.*[20] – Albert Einstein

When you discuss with particle physicists, be it about the data analysis, about the unsolved problems, about the way they communicate or whatever, sooner or later they will come up with the argument that so many people working in this field cannot err about the model. And I bet, on a psychological level (scientists are human beings after all), this is by far the strongest argument for every particle physicist if he is honest with him- or herself. Because it is just too convincing. Because imagining that tens of thousands of physicists have been investigating a flawed model of reality for generations, is out of the question. Terrible. Unthinkable. They cannot let the faintest suspicion of that kind come into their minds. And I bet, if you are honest with yourself, dear reader, the argument will have come to your mind, too. But it isn't sound. Not only it is dangerously circular, it is profoundly wrong.

The judgment about the standard model of particle physics is far from unequivocal among *physicists*. Many scientists intuitively feel that the high energy division is strange and not for them – I would say that the majority have at least some serious doubt about it. In spite of not being the experts, their opinion about the field is probably more substantiated than the detailed arguments of those brainwashed by their community. The intuition of physicists versus the expert opinions is what Daniel Kahnemann discusses in his bestseller *Thinking Fast and Slow*. His colleague Gerd Gigerenzer, director of the Max Planck Institute for Educational Research, *demonstrated* that housewives are better at predicting Wimbledon champions than tennis trainers, and taxi drivers' portfolios in a stock game outperform by far those of managers and consultants. Why? Because too much detailed junk information just throws dust in your eyes. Particle physicists are the prototypical geek experts who don't see the forest among all the trees. And although the average critical, open-minded, broadly educated physicist has a better (not only equally good) judgment than the expert, they will usually keep quiet. There is nothing to gain, other than being criticized, and he or she will not *feel* qualified to judge. Thus there is silence, and this is similar to a phenomenon that Nassim Taleb called "silent evidence." Some ancient sailors claimed that they had survived thunderstorms due to a religious talisman. Silent are those who are under the sea (though they had the same talisman). And there are those who survived without a talisman and felt no need to talk about – they correspond to today's real physicists, whom I would invite to express their opinions about the standard model and out themselves as doubters.

The Rubber Wall

When everyone thinks the same, nobody is thinking. – Walter Lippmann

However, it is anything but easy to express doubts, since the experts will usually interpret it as the lack of their sciolism. Thinking outside their box they call an "error" and if you don't follow their wacky assumptions about the laws of Nature you will be accused of "not understanding." *Their* understanding merely consists of rehashing how well the "predictions" of the standard model match the data. What you can predict is that the more staunchly a person declares

The Higgs Fake

this, the more clueless he or she will be. But it's hard to tell them to their face (I tried).

Particle physicists might be called experts, but mostly they are not even that. The detector specialist, the beam technician and the guy who calculates the decay probabilities of a given particle are far from understanding each other's field, let alone the theoretical stuff. Every experimenter would stutter when asked to explain how he sees those symmetry groups $U(1)$ x $SU(2)$ of the standard model in his data. They are condemned to parrot each other's opinions (of course, others outside the field also parrot.).

The whole system, unfortunately, almost entirely relies on confidence. Once you start asking questions, you usually get a polite though superficial answer, but if you follow up on something more closely, e.g. how to remove the background or how to handle radiation damping, sooner or later you hit the rubber-wall arguments "we do that very carefully," "lots of people have checked that," "that's been done for a long time" and so on. It's a line of defense that always works. They make you feel a little bit guilty in the same way religious people do – insinuating that it is your fault that you didn't waste part of your life dedicating your time to that nonsense.

But it is precisely here where you can distinguish the murky business of particle muddling from good physics. Ask a quantum optics postdoc and he will be able to explain to you how a laser works, talk with a geophysicist and you will get an idea how the Earth's magnetic field is generated, ask an astronomer how he estimates the age of a globular cluster of stars, and will you get satisfactory information. The better your preparation, the more easily and more deeply you will understand, but they would never tell you "It's been done reliably by many people, but you have to study in detail first until you can get a rough idea." Here is the key element of good science: If the researcher knows the details she also can give you a wholesale picture. If she has understood it, she can also explain it to Joe Sixpack. Ask particle physicists, and sooner or later they will quibble, resort to their authorities, say "we know," and ultimately tell you that you have to join their business to understand. Because they don't understand it either.

The Hearsay Factor

The last place on the planet where the "other people think so" argument can be used is science. – Nassim Taleb[21]

In questions of science the authority of a thousand is not worth the humble reasoning of a single individual. – Galileo Galilei

We parrot. We humans, unfortunately, are incredibly dependent on the opinion of our fellow human beings. This is a fact demonstrated by the famous conformity experiments of the psychologist Solomon Asch. A test subject misidentifies easily distinguishable visual patterns, just to comply with the deliberate wrong judgment of his peers.[22] Imagine how easily your opinion is influenced when the subject is difficult, you are yourself undecided, or you believe that other people have more expertise. Not only do we easily adopt opinions, but we easily fool ourselves into believing they are our own. And there is another striking parallel to religion: it is also the impressive technology, the pure size, cost and shared effort invested in the Large Hadron Collider that convinces people to believe in the standard model. Today's colliders have taken the role of medieval cathedrals. The harmonious singing in such a large building just cannot refer to a theoretical fantasy. Think about it.

From the beginning, big science has been the home of obedience and authority, profound enemies of science. There is no field that is dominated in a similar way by hierarchical structures (well, string theory, but I shall restrict my discussion to science). People like Samuel Ting or Carlo Rubbia, both Nobel laureates (in 1976 and 1984) ran their collaborations more in a style of dictatorship rather than with the kind of openness, fairness and culture of discussion that science needs. It is obvious that opinions in high energy physics are homogenized by social and hierarchical pressure.

The Emperor's New Clothes

There is no reason to assume that the structures present in the 1980s and documented in books such as *Nobel Dreams* by Gary Taubes, have changed in the meantime. Vigorous debate is much needed in science. But particle physicists don't discuss. They just *tell* everybody how much – internally, of course – discussion takes place in the community. Too bad that nobody outside the community ever noticed such debates. And, sorry, don't you have to write down an argument in a serious scientific question? But one never sees a divergent opinion published – if you think about from a historical perspective, that's really hilarious. Not even about one little detail of some detector calibration there is audible arguing. All the papers are the streamlined babble of several-thousand-author collaborations.

The lack of a culture of discussion in the particle physics community has rarely been more obvious than in the "seminar" held on July 4, 2012 in Geneva. ATLAS and CMS spokespeople Fabiola Giannotti and Joe Incandela both claimed that the signal had been measured with a significance of five sigma, a mathematically cocksure number that suggests the discovery was beyond any doubt. Though I didn't even put in doubt the statistical significance, a CERN physicist conceded in a panel discussion one year later in Berlin that he had considered the Higgs signal "weak" at that time. Why the hell didn't he stand up and say so? And was he the only one among 6000 collaboration members? Where were the others? Where was the discussion in that fake seminar?

At the end of the two talks, there was no comment, not even a slight addition, not a single question, let alone a critical one, just an endless sequence of subservient thanks, reminiscent of the Chinese People's Congress. Is this supposed to be science? All the evidence suggests that the big detector collaborations are uniform sociological groups, incapable of conducting genuine scientific debate, with individuals incapable of expressing dissent. They may occasionally talk about their doubts in private, but when it comes to make a stand they are muted, a textbook example of Solomon Asch's test subjects.

One of the most stupid arguments to justify the impenetrable analysis of its mega-experiments is that CERN "has a reputation to lose." Let alone the circular reasoning therein, it is this very mind-blocking idea that ultimately would forbid a concession of error. And they are damn afraid of that error. They are damn afraid of committing a small error. Even if you ask a technical question in an e-mail to a particle physicist, you often have to give an assurance that you are not going to quote him. And they are damn afraid of any debate about whether their enterprise makes sense at all. One high energy physicist wrote a four-page letter of complaint to the editor of a newspaper that had printed an interview with me, threatening that he and his colleagues would cancel their subscription if that happened again. But there was no word of response in public (though the editor had invited it). Such chickens make it easier for me to be outspoken.

The most cogent proof for the cowardice of particle physicists is their attitude towards string theory. String theory, for decades, has not made a single prediction that could have been tested at the LHC or elsewhere, and it is perfectly clear to every rational person that string theory has left the realm of science. Many researchers at CERN are convinced that strings are crap (e.g. the friends of the string critic Peter Woit who leaked the undisclosed Higgs results to him), but they pussyfoot around when it comes to making a stand in public. They fear too much that if a major part of theoretical physics is identified as fantasy, people will realize that high energy physics is just another stage for fairytale stories. Rather, CERN chief Rolf-Dieter Heuer gives talks at string conferences (and has recently humiliated himself by handing out the Fundamental Physics Prize to string celebrities), though he is perfectly aware that the only honest statement would be: "We can't measure what you guys calculate, goodbye." It is the ultimate convolution of two rotten branches which consider themselves still part of physics: one that predicts nothing, the other that measures anything.

Homo Sapiens Particellus

Convictions are more dangerous enemies of the truth than lies. – Friedrich Nietzsche, German philosopher

Along with string theorists, particle physicists consider themselves the elite of physics (of which evidence is provided by the money spent in the field). I don't buy that. Going into that business is incompatible with a deeper reflection on Nature as investment banking

The Higgs Fake

with charity. They have all become pain-insensitive with respect to the complication of the laws of Nature. To put it bluntly, particle physics is not the field where the intelligence clusters. It would be nice to compare the average IQ at CERN with the IQ of other physicists. But what is certain is that you won't find outliers to the top. A genius just doesn't go there. The smart people – see, for example, Ted Hänsch's autobiographical thoughts[23] at Nobelprize.org – were not able to put up with working in such an environment. They went to quantum optics, solid state physics, and all sorts of applied fields that flourished in the past decades. Most of these physicists felt that there was something wrong with the whole high energy business – but since physicists usually are nice people, they don't care much about spreading doubts about other fields.

Consider also the excessive timescale on which the last generations of colliders were built (high energy physicists deem this a sad factual constraint rather than a result of their decade-lasting giantism). Why should a bright young physicist, interested in the true laws of Nature rather than in pursuing his career, spend half of his professional life waiting for a Large Hadron Collider to be completed, to which he can contribute practically nothing on a fundamental level?

All this suggests that in high energy physics, for decades, an evolutionary selection of collective opinions, uncritical groupthinkers, skilled, narrow-minded mutants took place, while unorthodox thinkers, the independent mavericks, the individuals that rub you the wrong way, dropped out. No wonder their chorus is so uniform today. Whoever once enters the game will never be capable of nourishing serious doubts on the standard model, even if the evidence would suggest it. It is just too difficult, too poignant, too self-destructive, to say "my diploma is bullshit, my PhD is bullshit, all my papers are bullshit." A forbidden insight. And yet this is what a sober look at the whole enterprise involving thousands of physicists suggests.

It is hard to make somebody understand something when his income is based on not understanding it. – Upton Sinclair, American writer

Alexander Unzicker

Friends and Enemies

A friend of mine, a nuclear physicist, once advised me that criticism should always be constructive. The idea appeals to me, however, when talking about particle physics, I think it is out of place to be constructive. Literally, much too much has been constructed when building the higgledy-piggledy standard model. Yet, there is a simple mechanism why this occurs in science. If you work on a fashionable idea, it is very easy to make friends. You join a group or form a new one. If instead you challenge an established concept on which many people are working, they will be your enemies. There are no laurels to win in this fight. This is an inherent bias which necessarily leads to complication, once physicists in the 1930s had given up their self-control mechanisms that demanded parsimony in concepts. Not only are particle physicists blind to complication, but they don't even realize that the mechanism inevitably leads to it.

There are also friends and enemies among the experimental outcomes. Scientific communities tend to reject data that conflict with group commitments and, correspondingly, to adjust their experimental techniques and methods to "tune in" on phenomena consistent with those commitments.[24] It is common practice that results are withhold and reanalyzed when they do not conform to the expectations or show inconsistencies.[25] They are not even ashamed of it any more.

A decision in favor of particle physics must be more career-oriented, more based on good funding perspectives, more opportunistic, than the average creative scientist. The long-living traditions naturally are also based more on teacher-student relationships than in other fields. It is worthwhile here to observe that throughout the history of science, no really relevant discovery, let alone a scientific revolution, was made by a "student" of anybody who had been famous before. The geniuses are individuals, and the celebrities have always attracted sycophants more than creative minds. Just look at some research groups. The impact of a researcher usually diminishes greatly with every generation of doctoral adviser, thus being a scientific "grandson" of Heisenberg or Feynman isn't really something to boast about.

> *Every fool believes what his teachers tell him, and calls his credulity science or morality as confidently as his father called it divine revelation.* – George Bernard Shaw

Bad Money

> *We've got no money, so we've got to think.* – Ernest Rutherford

> *Partly because of the huge costs involved, a government contract becomes virtually a substitute for intellectual curiosity.* – Dwight D. Eisenhower

High energy physics, necessarily a part of big science, is well funded by its very nature, and this is another source of degeneration. In a way, money cements the stalling of theoretical physics, because the review process involving all these experts practically excludes the investigation of new ideas. Rather, axing an existing project means admitting a misinvestment and is correspondingly rare. This enhances the "friends and enemies" effect mentioned before. But also in the realm of huge experiments like the LHC, money may act as a form of intellectual corruption. One easily accepts the comfortable funding, while repressing the thoughts that should come to the mind of anyone with a sufficient intelligence when reflecting on the state of fundamental physics.

An important factor in the repression of critical thoughts is conferences. With their short-lived fashions, their excessive focus on the present, conferences are among the most counterproductive things in science. It is the very situation in a conference room, with several hundreds of scientists talking in all naturalness about "established" concepts (quark color, neutrino oscillations etc.) that psychologically forbids you to suspect that this concept might be nonsense. Conferences are intimidating brainwashing ceremonies forging clean collective opinions. They have no use at all besides the small side-benefit of providing cheap vacations for researchers. No discovery was ever achieved as a consequence of a conference. If ever, the one-to-one conversations of Heisenberg, Bohr, Einstein and Schrödinger advanced physics. Not only is much of fundamental physics wrong, but the way it's done today is also wrong. My book inevitably will lead to people accusing me of being dismissive about scientists who are honestly committed to their projects. I do not deny that these people exist, and it is not my intention to insult them. However, there is no right to be fed unless you are working to find out something relevant. It may be that many believe this, but when looking soberly at the efficiency of their endeavors in the past hundred years, I think it's very hard for physicists to be honest to themselves.

To sweep this under the rug is dangerous, because sooner or later the public will want to know what we have learned from the particle zoo produced in the colliders – apart from the question of its usefulness. This question about utility is something we really have to fear – because it would mark the death of curiosity, the death of science.

Thus I definitively want money to be spent on fundamental questions (rather than feeding a perverted financial industry), but one must confess that at the moment the money is wasted.

The Higgs Fake

CHAPTER 6

LACK OF TRANSPARENCY

WHY NOBODY CAN OVERSEE THAT BUSINESS

Michael Faraday, the legendary discoverer of electromagnetic induction, conducted about 10,000 experiments in his life, all of them meticulously documented in his lab notes. At the Large Hadron Collider, about 10,000 people are conducting *one* experiment. Go figure if this is the same kind of science as 200 years ago. The scientists involved don't know each other, and they know even less what everybody else is doing. The sheer number of people working in *big science* creates its inherent problems (Alvin Weinberg and Derek de Solla Price have written insightful books about this), and CERN's collaborations are the biggest ever. It is impossible that anybody can maintain an overview about how the experiment actually works.

The technical details have an unprecedented complexity, the amount of data is overwhelming, the underlying standard model has grown to an unbelievable complication. The analysis as a whole has become impenetrable. Sometimes particle physicists, to justify their muddling along, say that "we need a new Einstein" (who descends from heaven and comes to the rescue) for a better theory. This is a profoundly stupid argument, ignoring the fact that Einstein in his day was up to date not only in theoretical physics (relativity, quantum physics, thermodynamics), but also in all the experimental results backing the theories. It's not that unlikely that there are physicists with comparable capabilities among the world's seven billion people. But an Einstein cannot exist because it is now absolutely impossible to oversee "fundamental" physics. No way. Here again, the question is: Did humans become considerably dumber in the last century, or have we messed up physics to a state that is beyond remedy? We don't really observe swarm intelligence among humans. And the last place we would find it is CERN.

Some thirty years ago, Pierre Darriulat, the head of one of the two big CERN experiments, UA2, commented on the complexity: "In fact, outside UA2 and UA1 physicists simply could no longer understand the two experiments. They were too big and too complicated."[26] This was already an alarming sign of sickness. And there is certainly no reason to expect that things have improved with the Large Hadron Collider delivering 1000 times more data that requires ten times as many scientists to run and analyze it. Particle physics has become swamped by these ever-increasing data shitting machines.

Just ask anyone involved in the experiment in a little more detail about the background reduction, radiation damping (a basic, unsolved problem!), and you soon arrive at the point where they assure you that "all this has been done properly," "you easily underestimate the complexity of the task," "you need to understand the detector," "many people have checked this." Everybody relies on each other, but no one really understands the experiment in its entirety.

The Devil Is in the Detail

If we look at the "lowest level" of raw data, the unprocessed detector responses, there is already an overwhelming complexity. There are probably lots of cases where our understanding of the material properties is insufficient, at least to justify the level of precision that is claimed for the final outcome. Surely, scientists try to do their best, but it is naïve to believe all possible artifacts, unknown interactions between effects, etc. can be properly controlled. None of the detector components are understood at the fundamental level (this is no criticism) in the sense that their response can be calculated from first principles. They all have to be tested, calibrated, and a lot of simplifying assumptions enter here. I don't deny all this is done reasonably and to the best of the existing knowledge, but still some unexpected, overlooked or unknown material effect could greatly influence the outcome. Just think about the badly soldered electric contact that damaged the experiment in 2008. That was actually visible, but a technical defect just affecting the analysis may be not.

Processing the data with computer programs is another serious source of possible errors. Surely, the task is not less complex than programming a operating system, which, as we know, is accompanied by a continuous production of bugs and patches. The difference is that operating systems are continuously strengthened by hacker activity and the antidotes to it, while there is no motivation to correct errors unless they manifest in obviously wrong results. I doubt that the millions of lines of computer code are free of critical errors. The modelers would deny this, but what they cannot deny is that the whole thing is non-transparent. Though the programs are in theory undisclosed within the collaboration, there is no incentive for anybody to scrutinize their colleagues' analysis. True transparency would require open, commented upon programs.

Still, everything might work correctly at the computational level. But the lack of transparency also involves the underlying physics. In such an arbitrary and complicated theoretical model, nobody can thoroughly check the consequences if something is wrong. Thus it becomes almost impossible to realize that built-in mechanism fooling the community when every fifteen years a still rarer signal is declared as a discovery.

Looking, Blinding and Blinders

There is no transparency either with regard to the decision making of collaborations like ATLAS and CMS. And this also contributes to the inevitable bias towards the discovery. In one of the few intelligent articles on the Higgs discovery, the authors ask: "One team, say ATLAS, discovers the particle while the other does not. Who would get the credit, the acclaim, the prizes?"[27] However, it is essentially the same dilemma that has been pointed out by Andrew Pickering. The benefit of doubt goes to the discovery.

The meticulous work of scrutinizing, considering one's own errors in the analysis, all this implicitly puts in doubt one's own competence and is an unconsciously frustrating thing, while the expectation of the long sought-after breakthrough, imagining being the first, is incredibly exciting. There is no way to for humans to escape this dilemma, and it is greatly enhanced by the lack of transparency. The bias has sunk in so deeply in the daily practice that it isn't perceived as such any more. When an analysis delivers the expected results, there is no problem, it usually gets published. If there is a deviation instead, it is practice to delay the publication, without revealing the nature of the anomaly, and without making public how the mistake had occurred. Or just nobody talks about the deviation.[1] Usually it is objected that things have to be discussed first inside the collaboration. But this is not science. It is rather reminiscent of the discussions of politbureaus in the Cold War era.

Another counterargument favored by high energy physicists is that they do "blinding," that is, a part of the information of the data is withheld, to avoid placebo-like effects of the sort that occur in medicine. These are necessary and neat techniques, but (since they do not put into question any of the field's prejudices) far from helpful to avoid the large-scale self-deception that their analysis makes sense. As in many high energy physics techniques, there is no thought-out methodological basis for "blinding" or "triggering," all have grown up out of necessity in a spontaneous manner. They suffer not only from ad-hoc concepts, but also from an ad-hoc methodology. Instead of patches like "blinding," an absolute transparency in processing the data is needed.

Transparency means publication. Publication, as elsewhere in science, is currently entangled with evaluation of some kind, which leads

---

[1] I may be mistaken, but just look at https://atlas.web.cern.ch/Atlas/GROUPS/PHYSICS/CONFNOTES/ATLAS-CONF-2013-067/. The results seem to heaviliy disagree with the standard model for higher energies.

The Higgs Fake

to the suppression of opinions that would endanger the sacred cows of an established field. This is a considerable problem we cannot deal with in detail here. However, the weaknesses of peer-review are more and more discussed. Stuff like indices, impact factors, all that rated garbage, has grown because there is no real measure of quality for physicists (and in fundamental physics, no real quality either).Nobody cared about the impact factor of Einstein or Dirac, and likely, a posthumous count of Ptolemy's index would be impressive. Scientific revolutions (which necessarily occur once in a while) don't originate from papers that ride on the fashions that impress reviewers. Today's publishing in the fields dominated by the big collaborations is often a rehash without any ideas. To come to particle physics in particular, I wonder what it does mean when a paper with 3000 authors goes to a review process in which the validity of the results are supposed to be checked. Can anyone explain how that works? With several hundred reviewers may be?

This already highlights the absurdities. What bothers me more is that, when you read the LHC results, they are usually published at the top level, e.g. discussing which particles could be interpreted as having decayed from a Higgs boson.[28] But there are multiple layers of assumptions beneath, and the more one goes down and approaches the raw data, the sparser the publications become. Here again we encounter the phenomenon that it is much more pleasing to deal with the big questions rather that to dig into the dirty business of detector components, their difficult response behavior, calibration problems and so on.

Whatever you might believe about the validity of the data analysis, let's keep in mind that publication is essential for reproducibility. That's the goal, that's why Faraday wrote down his lab notes.

Reproducibility – What Does That Mean?

Reproducibility is at the heart of science, the key element of scientific methodology, the thing that distinguishes physics from spoon bending, astrology and all kind of snake oil activities. You have to be able to repeat something, in all places, at all times, with different people in different laboratories. Particle physics has a problem with this. CERN is an impressive lab, but for a gigantic particle accelerator the world can only afford to build once, reproducibility is not that easy. Sometimes it is claimed that "checks and balances" are realized with competing groups like ATLAS and CMS. But not only could artifacts of the proton beam well affect both groups, more importantly, they all use the same software to distinguish signal from background – and this is where the wrong concepts are buried. You probably don't realize how much theory is hidden in these "experiments" and all this, bear in mind, is practically secret. The competing groups would never put in doubt the long established wisdoms of the community, such as that the W boson is a reasonable concept, and there isn't anybody else who could. To claim that ATLAS and CMS guarantee an independent analysis is like saying that the healthy competition between McDonalds and Burger King will necessarily result in a healthier diet for their customers.

Besides that, the "independence" is merely a fig leaf. There are people in ATLAS who have a romantic partner working at CMS. Who could control, let alone sanction whether they talk about their results? It is also well known that the respective collaborations have always looked over another's shoulder, as happened in the 1980s when UA2 hacked UA1 computers (and likely, vice versa).[29] And I would bet that Rolf-Dieter Heuer, before introducing the seminar speakers on July 4, 2012, knew very well that they would not contradict each other. It is documented that his predecessor Herwig Schopper assured himself many times of avoiding potential embarrassment during the announcement of the W and Z bosons in 1983. To talk about an independent analysis is somewhere between believing in Santa Claus and telling an explicit lie.

Even if it worked (it does not), reproducibility means little more than two groups checking their results of the day. What is needed is sustainability. How should one repeat the experiments after, say, 50 or 100 years, when there is no longer direct access to the results? Take that couple of publications and just build it again? And what about the millions of lines of computer code that are essential for the analysis? For all this there is no documentation that would merit the name, because the whole business is a present-focused, funding-orientated and arrogantly ahistorical enterprise that prefers the scientific quick buck to a thorough investigation of the laws of Nature. Why should anyone care for a longer period than his funding award, let alone for longer than his career? Particle physics has reached a state of nonfalsifiability, nontestability, inscrutableness – a paradise for ideology to flourish.

If the experiment itself is not easily repeated, it is at least the bloody business of the people involved to conserve the data properly. But not even such an obvious logic seems to be observed. The data of the Tevatron collider, which was shut down in 2011, has not been maintained due to funding problems.[30] To save a couple of thousands of dollars for the busy present, the long-term results of billions of dollars are squandered. People should rally against it.

A Truly Independent Test

There is one type of reproducibility, aside from the usual one, that constitutes an impeccable test of a scientific theory, one that establishes evidence without any reasonable doubt (that means except conspiracy theorists, creationists or other irrational folks): Application. Beware, I am not advocating that research has to be restricted to the narrow limits of foreseeable technology, that would be the death of science. But in history, the true discoveries sooner or later have always transformed into technology. Why shouldn't they? That you can't plan technological revolutions is one thing. But too little attention is given to the undeniable evidence provided by application. Darwin laid the foundation for genetics, Mendelejev's periodic table for all the chemical substances synthesized afterwards. Hertz's electromagnetic waves led to today's wireless communication world. Apart from the commodities, the essential thing for science is that the phenomenon is repeated numerous times and tested by these applications, thus you could now call anyone a crank who denies the existence of electromagnetic waves. And in the same way, the time dilation in gravitational fields is tested by the atomic clocks in GPS satellites, and quantum effects are tested in digital cameras, lasers,

NMR medical imaging, and occasional Fukushima shows that there is something that works with nuclear physics (though it's not their safety policies). Such evidence is omnipresent and beyond any doubt, not least because producers of digital chips are anything but interested in testing a theory by Einstein or Bohr. History just tells us that revolutionary discoveries, voluntarily or not, transform into something useful. On the other hand, none the findings of elementary particle physics have been tested outside the academic environment. There is obviously nothing you could build with a Higgs or W boson or with a top quark, let alone sell. A lifetime of some 10^{-25} s is kind of short. But also the established particles, like pions, nobody has used in another environment. Muons are an exception, because besides their original detection in the realm of nuclear physics, quantum optics experiments were able to reveal their masses independently (the question remains why they are so similar to electrons). And not even neutrinos, according to the standard models an inevitable consequence of every nuclear reaction, have been able to prove their use outside the neutrino physics community – though in principle, one could think of a neutrino-based apparatus that detects nuclear weapon tests or even atomic submarines.[1] Frankly, I don't have any idea what a Higgs boson could be useful for. Researchers will now cheekily retort and compare their forced discovery to Hertz's electromagnetic waves, but they know pretty well it's nonsense.

The Academic Cuisine, to be Restructured

It is that lack of external control which makes me suspicious of the findings of particle physics. If you construct an airplane battery badly, it will catch fire, a poorly understood laser will not weld, if you have a bug in your software, it will cause your satellite to crash or your train to break, or whatever. In particle physics, there is no obvious consequence: rather, theoreticians – since they don't do quantitative predictions themselves – can always digest the data and trans-

---

[1] I wonder by the way why neutrino detectors investigating Japanese nuclear plants have not determined the precise hour of the Fukushima meltdown with their data.

The Higgs Fake

form into some unexpected discovery. Particle physics, for decades, has had nothing to produce, nothing to deliver, nothing to get to work outside academia. CERN is the biggest company in the world that defines for itself the value of its output. Let a pilot fly his construction, let a financial product trade on the market, all these are healthy tests. But there is no way, to speak with Nassim Taleb,[31] to make particle physicists eat their own cooking.

If we look at the menu of data analysis, it seems that everything is devoured at once instead of as a sequence of dishes. The basic problem is the all-in-one, "vertical" analysis performed by the respective groups. From the very first stage, triggering (data selection), calibration, simulation, background removal, interpretation as particles, all this is done in a single process and under the supervision of one group, with no intermediate results published. And if at the end, the energy of the signal disagrees (e.g. the ATLAS and CMS values of the Higgs mass from 124.3 to 126.8 GeV), then one goes back to the bottom of the analysis, the very first stage and tinkers with the calibration to get the energy fixed. Instead, what is needed is a "horizontal" structure in the analysis where the first layer of elementary analysis is published, without anyone being interested in the final result. To see how things could be done reasonably, have a look at the Sloan Digital Sky Survey (SDSS) astrophysical data which are available online. There is one so to speak primitive stage of the analysis where the CCD signals are transformed into brightness of a certain color. Most people won't be interested too much in how this is done, but if you want, you can look up the public code. Then, at a second stage, the brightness is assigned to a color filter of a certain galaxy, and so on. If you don't like the algorithm, just make your own. This is freedom of research. Finally, you can deduce from the color information the cosmological redshift of the galaxy, and if you do it for a large sample of galaxies, you may arrive at the rate of expansion of the universe. If you believe in another cosmological model, it's up to you to interpret the data in these terms and see if it is convincing. Collider data analysts just offer a sink or swim. There is no way to get out at an intermediate stage. It is a death sentence for transparency.

Alexander Unzicker

Against Legend

It is untenable that the definition of a W boson should be hidden in an undisclosed, and probably impenetrable, computer code. This is, methodologically, the most absurd aspect of the particle physics enterprise: There is no way for an alternative, possibly reasonable, theory to make sense of the data. Just to read it, you have to believe all the crap of W bosons, top quarks and gluons. What is still not there (and very far away), is a completely model-independent formulation of results. (If one could make any sense out of the amount of data is another question.) Practically no one has access to a simple form of data, such as how much energy was deposited at a given time and detector point.[1] Particle physicists aren't able to generate that kind of data in a useful way. If you ask them about it, they come up with stupid excuses such as you have to know the detector, you have to know the calibration which is complicated, bla bla, and it wouldn't be useful for you unless you are an expert. Correct. But it is damn well their duty to deliver the data in a useful, accessible form. It is their duty to the taxpayer, to the public, to the interested researcher, and to humankind who in later generations may want to follow up the big questions outside the incestuous circles of high energy physicists.

Particle physicists, confronted with the demand that data paid for by the public should be in the public domain, usually start moaning that the amount of data is overwhelming, "you don't have any idea how complicated it is," "we can barely handle it ourselves," "get funding for new computers first before you start talking," and similar bullshit. Sorry, but who got us into this mess if not you guys? Whose fault is it? Today we are facing the problems of a 60-year long gigantism of particle physics that thoughtlessly expanded towards ever higher energies, higher luminosities and ever more data. And now, this very thoughtlessness is taken as an excuse for no longer being able to work with clean methods, to be reproducible, to do science.

---

[1] Called minimum bias data, or, a little more advanced level, common n-tuples.

The Higgs Fake

It's like a guy weighing 200 kilograms having gorged himself for decades who complains that he is too fat to move.

Is the LHC data too complex to be put on the internet? Well, why not deal with smaller bits? Take care of the Tevatron data instead of letting it wither. Put the DESY data on the internet, and the data from which the discovery of W bosons was deduced in the early 1980s (SPS and LEP). If it was the revolution it was claimed to be, let the world participate! Publish the raw data of the 1974 Gargamelle experiment which "established" the weak neutral currents for which Weinberg got his Nobel. And I'd like to see Lederman's famous muon-neutrino experiment from 1962 repeated with public data, and Douglas Hofstadter's scattering experiments from 1961. Wouldn't that be at least showing respect for those iconic figures in particle physics? But there is nothing, no data, no documentation, not a single wire chamber photograph, not a single collision publicly available of which you could make sense. And this is a scandal.

Particle physics behaves as if astronomers were to store their galaxy photographs in a filing cabinet. The scandal has an ironic note, since the World Wide Web, the tool uniquely apt to distribute such information, was invented at CERN. It has changed our social behavior, it has created economical and informational superpowers like Amazon and Google (and their big brother), it has catalyzed political overthrows, it has revolutionized the world, except one place: particle physics at CERN, Geneva.

PART II

HOW IT CAME ALL ABOUT

CHAPTER 7

THE STEADY DEGRADATION OF THOUGHT

FROM POORLY UNDERSTOOD PHYSICS TO HIGH-TECH SPORTS

> *Once the problem is eliminated by an excuse, there is no need to reflect upon it any more.* – Erwin Schrödinger

Elementary objects seem to behave sometimes as particles and sometimes as waves. At the legendary Solvay conference in 1927 almost all the famous physicists of the day came together to discuss this paradox, but the revolutionary findings of quantum mechanics remain puzzling even today. Werner Heisenberg, Wolfgang Pauli and Niels Bohr advocated the "Copenhagen" interpretation of the waves expressing a probability of finding a particle in a given place. Einstein,, often quoted as saying, "God doesn't play dice," fought ferociously against any probabilistic element in Nature, and Erwin Schrödinger accused the Copenhagen interpretation of denying objective reality, an assumption he considered "foolish." Heisenberg, assisted by Pauli, declared the problem resolved and saw himself as the winner of the debate, provoking an angry reaction by Schrödinger: "The Göttingen folks are abusing my beautiful wave mechanics to calculate their shitty matrix elements." Pauli instead mocked "Schrödinger's childish attempts to get rid of the statistical interpretation of his wave function." These little anecdotes show a grave, if not tragic, development: The founding fathers of quantum mechanics, the most important discovery of the twentieth century, disagreed severely on how to make sense of their common offspring. The 1927 conference was the last intense debate among all leading physicists, and a fragmentation of physics disciplines followed that was, likely, the beginning of its crisis.

Out of Simplicity

Several events in this period marked the paradigm shift that was to come. Until 1930, atomic physics made do with just two particles, the heavy proton that constituted the nucleus and the lightweight electron that formed the atomic shell. The earlier discovery of radioactivity, however, would soon lead to serious trouble for this simple model. At the same time, severe problems resisted a solution: The mass ratio of proton to electron, a mysterious value 1836.15..., is an unexplained number. Likewise, nobody can calculate the fine structure constant $1/137.0359...$, a combination of constants of Nature. The attempts to generalize quantum theory to electrodynamics were not straightforward; they rather worked around problems rather than getting to the bottom. The basic problem mentioned in chapter 2 persisted: Classical electrodynamics is inconsistent, since its equations would assign an infinite energy, hence an infinite mass, to the electron.

"Renormalization," a theory developed in the 1940s, now says that this infinite mass of the electric field is just about compensated by an infinite negative energy (never seen that before) of a "naked" electron. Too much for you to understand? For Dirac, too. He said, "This is just not sensible mathematics. Sensible mathematics involves neglecting a quantity when it turns out to be small – not neglecting it just because it is infinitely great and you do not want it!" The naked electron, dressed with the electric field, nicely describes the observations. However, the problem of inconsistency is swept under the rug. Much later, even Feynman doubted the validity of renormalization: "It's surprising that the theory still hasn't been proved self-consistent one way or the other by now; I suspect that renormalization is not mathematically legitimate." The gradual takeover of that concept, seen in a broader context, contributed to the transition from philosophy-based physics to the technical, math-recipe form that had become fashionable in the late 1920s. Paul Dirac, in an early intuitive statement, was particularly skeptical towards the "complicated and ugly" theory called quantum electrodynamics (QED):[32]

> *Some physicists may be happy to have a set of working rules leading to results in agreement with observations. They may think that this is the goal of physics. But it is not enough. One wants to understand how Nature works.*

Dirac might have been the theoretical physicist with the most exceptional skills, but after years of frustration he lost interest in quantum electrodynamics, because he felt that no progress was achieved. As his biographer Helge Kragh notes[33] Bohr, Dirac, Pauli, Heisenberg, Born, Oppenheimer, Peierls and Fock came to the conclusion, each in his own way, that the failure of quantum electrodynamics at high energies would require a revolutionary break with current theory.

Grit your Teeth and Get to it!

> *Quantum electrodynamics is a complete departure from logic. It changes the entire character of the theory.* – Paul Dirac

A technical, not to say superficial, way of doing physics gained ground, leading to a complete reorientation of physics in those days. To be blunt, it was at that point that, due to the lack of true understanding, the collective displacement activity in the form of feasible but unreflective calculations began to take over. Kragh continues:[34]

> *With the recognition of new particles [...], the existing theory – gradually*

improved in its details but not changed in essence – proved to be quite workable after all. The empirical disagreements became less serious, and by the end of the thirties most of the young theorists had learned to live with the theory. They adapted themselves to the new situation without caring too much about the theory's lack of formal consistency and conceptual clarity. [...] When the modern theory of renormalization was established after the war, the majority of physicists agreed that everything was fine and the long-awaited revolution was unnecessary.

Kragh's comment precisely identifies the beginning of the sickness that became today's intellectual epidemic of particle physics. Back then, they blew it.

Soon after the war, Richard Feynman and others developed the theory of quantum electrodynamics further, achieving a remarkable precision with tests like the Lamb shift or the anomalous magnetic moment of the electron.[1] Independently from both Heisenberg and Schrödinger's approach, Feynman brilliantly reinvented quantum mechanics. He is a character you simply cannot dislike. His textbooks are full of deep thoughts, and despite his genius he does not try to shine with brilliant calculations that are often all smoke and mirrors. Instead, he honestly admits when something is not understood. His contributions are outstanding, yet the role he has taken in modern physics has become somehow frightening.

The Paradigm Changer

Feynman's era started when physics had to recover from the lack of orientation caused by the quantum revolution. His ingenious, powerful and optimistic character was not inclined to delve into the boring arguments of the Copenhagen interpretation and its opponents, rather his idiosyncratic way of thinking made a hit with quantum electrodynamics. Characteristically, Feynman didn't try to hide the mathematical shortcomings of the theory, and at the same time he was thrilled about its success. He lived on the sunny side of phys-

---

[1] The g-2 experiment is spectacular, but the theoretical accuracy is usually quoted to an exaggerated degree. Here also, public computer code would be desirable to repeat the calculations.

ics, unlike musing thinkers like Einstein, Mach or Dirac. Today, when the sunny days have given way to the fog of rampant complication, it comes back to roost that Feynman did not take too seriously some thoughts of Einstein, Dirac and Schrödinger (as you can see from his occasional sloppy remarks). It had been tempting for Feynman not to bother with the burden of the old unanswered questions, but physics became too lightweight with him. Now, as the theories are running from one dead end into the next, we should remember Feynman's words: "Whenever we encounter a mess of too many problems, it's because we stick to established methods." Sadly, that's true also for the methods the whole world adapted from Feynman: those of quantum electrodynamics.

The Fine Weather Theory

A modern, though weird, variation of quantum electrodynamics is quantum *chromo*dynamics (QCD). Many people were enthusiastic about the "wonderful similarity" of the theories, but it was Richard Feynman himself who mocked such enthusiasm: "it's not because Nature is really similar; it's because the physicists have only been able to think of the same damn thing, over and over again."[35] While quantum electrodynamics delivers some very precise predictions at least, the corresponding results of QCD are poor.[36]

Yet there is a more severe conceptual problem in QCD, which has its roots in the superior QED. Freeman Dyson, a mathematically skilled colleague of Feynman, demonstrated[37] that the usual approximations (a very familiar technique also in engineering) do *not* converge the precise physical values. Since electrodynamics has a small "coupling constant" (actually the tantalizing number 1/137), the consequences are not that grave, but with the much greater value of the strong (nuclear) interaction, it turns out that the method – for *purely mathematical reasons* – cannot even produce an approximation of say, 20 percent. Go figure. This, considering it soberly, is a bizarre situation. A physical theory, designed to determine the true characteristics of Nature, allows itself to deviate from them in an arbitrary way. At this point, I wouldn't give a damn about it. Quantum chromodynamics is a paper tiger that by construction is unable to deliver reasonable results. The absurdity is usually wrapped in the term "perturbative methods don't work here" and one needs "nonperturbative methods" (which is something like dry water, or like paying by assuming you have money). But nobody rejects the whole thing as nonsense. It has been the leading theoretical paradigm for six decades.

Elusiveness

Let us return to the experimental situation in the 1930s. Though the old problems were still pending, physics was now prepared to accept a series of particles, something that had been unthinkable in a simplicity-oriented search for the fundamental laws of Nature. Within ten years, the positron (the positively charged twin or antiparticle of the electron), the neutron (an electrically neutral component of the nucleus, formerly believed to be a proton-electron composite), and the muon (a copy of the electron, though 206 times heavier for unknown reasons) were discovered. There is no doubt about the physical reality of these particles, but a cogent reason for the existence of such a variety was missing, all the more an explanation of the masses. Feynman, for example, later speculated that the muon could be just an excited electron, but he couldn't make a coherent theory out of that – such theories would be needed however, if physicists had not abandoned their desire for understanding.

In this climate, physicists were no longer reluctant to accept the neutrino, a particle that had been postulated by Wolfgang Pauli in 1930. It helped to account for the missing energy in the beta radioactivity, where the electrons expelled by the nucleus were just too slow. Pauli was aware of the methodological danger and wrote, right after his proposal: "Today I have done something that a theoretical physicist should never have done. I replaced something we don't understand with something we can't measure." However, the neutrino suddenly took on a life of its own. Later experiments have provided evidence, but there is still something wrong with neutrinos. I know that some people will shake their heads and recite the numerous neutrino experiments. I am not going to deny that there is a physical reality in the phenomena signals coming from nuclear fusion or fission reactions, but let us pause for a moment and see where this story has led in the meantime, even if we have to anticipate a little. It took until 1956 to identify a very tiny signal close to a reactor as neutrino-induced.[1] Soon afterwards, however, a series of unexpected and contradictory effects called for two more types of neutrinos. Not enough here, the three types altogether are believed today to transform into each other by so-called neutrino oscillations. Since the primary goal here is high energy physics, I cannot go into the details of the papers published by large neutrino collaborations of hundreds of scientists[38] (my general comments apply though). It is however clear that neutrino physicists, in the long term, follow the same mechanism of fooling themselves with ever more complicated assumptions about what they don't understand. The recent developments in the field are plainly ridiculous. For example, let us see what happened at a leading neutrino conference[39] in 2011. Again, the data could not be explained by the existing types of neutrinos, and instead of[II] a dawning insight that the whole system is crap, they put their next free parameters into play: one or two "sterile" neutrinos (meaning, with absurd properties), and a "non-standard-interaction" (whatever that means). As a researcher from Yale University pointed out elatedly, all these phenomena *beyond the standard model* (meaning, not understood) are *very exciting*, and if after a series of unjustified auxiliary assumptions you still can't make sense out of your data (which is called *new physics*), it's *fun*.

Neutrino physics is sick. The problem, of course, is not a single deficient talk but the intellectual degradation of 300 people in a conference room applauding such obvious methodological nonsense.

---

[1] A very ingenious and at the time costly experiment, that should be permanently repeated due to its outstanding relevance, as I shall argue later.

[II] Some cosmologists went out on a limb in saying that their data – right after three species had been established – suggested three species of neutrinos. There is no doubt that these guys will be able to fine-tune the number to four, five or six as well.

The Inevitable Human Factor

We have deviated a little into neutrino physics, but obviously there is pretty much universality in the way humans do science. Thus let us judge the above situation a little less harshly. It is understandable that a surprising phenomenon is cured by some new theoretical hypothesis, but the sequence of small, seemingly reasonable steps is often a path in the wrong direction. Science progresses here in a strictly local and relative manner, each step being only slightly more exotic than the previous one. The sociological fact seems to be however that scientists are blind to the absolute amount of noncredibility, when many layers of just tiny absurdities had been approved by their respected senior colleagues.

Instead, one could wonder about the law of energy conservation during nuclear decay processes. Bohr thought the revolution he considered necessary had to go along these lines, but this probably would have required a single, ingenious leap away from conventional thinking. Needless to say, today's parroters will come up with an arrogant smirk when you talk about Bohr's idea: "We understand all this much better today."

With the establishment of the neutral particles neutron and the neutrino, physics started to suffer from the problem of cleanly classifying its results. Electric neutrality means invisibility to most instruments, and this is where theoretical assumptions necessarily enter the analysis. This principal problem has been amplified with today's multitude of neutral particles (the collider geeks don't realize it at all), but the difficulty of separating neutron signals persists in experiments![40]

In 1950, another particle, the pion, was detected in the cosmic radiation. It cemented another unsatisfactory idea that became the basis of the later standard model: the four interactions. But researchers could continue to work with the superficial classification, assigning further particles to them. For the leading physicists of the 1920s it was perfectly clear that such a variety of four alleged forces was the product of poor understanding of electromagnetism rather than something fundamental. Bohr, Einstein, Schrödinger and Dirac never squandered a thought on such patchwork. They died lonesome.

Alexander Unzicker

Industrial Botany

Meanwhile, the center of what was considered cutting-edge physics had shifted elsewhere. Accelerator physics, powerful due to the development of nuclear weapons and generously funded, became the leading discipline. A new type of practical, ambitious, optimistic, skilled and team-oriented researcher emerged. The reluctance of the old-fashioned physicists towards theoretical complication faded away, and everybody competed in the race for new particles. Thus a rather brainless production took place in the 1950s with the new machines, illustrated by an only slightly ironic quote by Patrick Blackett, who had won the Nobel for a detection technique: "We really try not to discover more than one particle in one paper." Little reflection hindered the regression into data collecting, and Enrico Fermi was one of the few who felt uneasy about all the particles: "If I could remember all the names of these particles, I would have become a botanist." Louis Alvarez, another Nobel laureate (1968), was explicitly lauded for detecting a large number of new particles by the Nobel committee, which referred to his "decisive" contributions to particle physics. Physics, however, had decided to mushroom. In 1951 the number of particles was about 15, eight years later it had doubled, and in 1964 there were around 75 particles.

We should pause here for a moment and ask what "particle" actually means. When the beam of an accelerator hits the target, a variety of reactions occur, with their respective probabilities (usually denoted as "cross section") that can be plotted. To put it bluntly, every little bump in the diagram of cross sections can be interpreted as a particle – something happens there, so to speak, we have no clue what precisely, but let's call it a particle. The narrower the peak emerging from the background, the happier people are, because this indicates a relatively long lifetime of the "particle". However, physicists in that period started to classify particles regardless of their lifetimes – which were incredibly small sometimes, such as some 10^{-25} seconds for the delta particle. This is methodologically absurd, because there is no way to get out of the collision point into any detector – the interactions of such particles remain totally theory-inferred.

The Higgs Fake

The Slippery Slope

To summarize, the key elements that marked the turning point of physics were: the acceptance of several elementary particles, or better, accepting that physics could progress with an increasing number of particles (rather than being considered as a lack of understanding), the establishing of four interactions, the dismissal of Einstein's solitary attempts to unify electromagnetism and gravity; moreover, an utter separation of gravity from the rest of physics (in fact, there were no experiments to do, no habitat for a theoretical-observational symbiosis). All this is not just a sequence of events, but something that changed the character of physics – it is the moment when it metamorphosed from the search for Nature's deep questions to a hightech sport with some physics-related detector equipment.

I cannot spare you this historical picture; there is no way to evaluate the findings of a branch of science unless one contemplates its development. And it was that period until 1960 in which the principal absurdities of particle physics were established: nonsensically short lifetimes, postulation of a huge amount of neutral particles, lack of quantitative results, the unlimited increase of free parameters. In summary, people became blind to the methodological pitfalls from which physics now suffers.

Alexander Unzicker

CHAPTER 8

THE TAKEOFF TO METAPHOR

BRUTE FORCE FINDING, BRUTE FORCE EXPLAINING

> *If we are uncritical we shall always find what we want: we shall look for, and find, confirmations, and we shall look away from, and not see, whatever might be dangerous to our pet theories.* — Karl Popper

In the last chapter we saw a science that once dealt with the elementary building blocks of Nature devolving into an accelerator industry that produced an incomprehensible bulk of particles. What followed were attempts to squeeze the unintelligible into a descriptive picture. This required new theoretical concepts, which at the same time eased the crossover from a graspable picture of particles to unintuitive, merely metaphoric objects.

The basic experimental setup of particle physics never underwent great changes. In 1914, Ernest Rutherford discovered the atomic nucleus by gunning it with fast alpha-particles that occasionally were deflected ("scattered") from their original direction. This "elastic scattering" of projectiles then became a common method to investigate the structure of elementary particles. By measuring the angle of deflection of the projectile (often electrons) one could calculate the structure of the target (the nucleus) from it. Robert Hofstadter received the Nobel Prize in 1961 for a precise determination of the charge distribution of the proton (which is responsible for the deflection) called "form factor." However, all this does not mean we understand the proton, because "charge distribution" is an entirely classical concept that Erwin Schrödinger had already tried unsuccessfully to preserve. No wonder that measurements continue to run into contradictions[41] – it is simply the old quantum mechanics that tells us we have the wrong concepts. Of course, that doesn't prevent physicists from indulging in their displacement activities, for which new ideas are not necessary.

Highly Energetic, Not Highly Enlightening

In the mid 1960s, the form factors seemed to become a little boring to the physicists Kendall, Friedmann and Taylor, and they came up with a truly novel idea you would never guess: one could use higher energies. The so-called *inelastic* scattering, distinct from pure deflections, would produce new particles from the excess energy, quite a lot of them actually. David Lindley, former editor of *Nature*, commented:

> *The standard wisdom at the time was that the mess of debris created by inelastic scattering was too complex to be understood in a way that would shed light on the inner structure, if there were any, of the proton. Inelastic scattering was considered as highly speculative at best, as waste of beam time at worst.*[42]

While Hofstadter quit the program for that reason, Kendall, Friedmann and Taylor found a solution for the impenetrable results: not to look too closely. While you might think that a clean analysis of data requires the identification of the entire outcome, they restricted their interest to the question whether any production was under way. With this rectified point of view it was then easier to do a little statistics. The fact that *some* regularity popped up *somewhere* is not exactly surprising – otherwise the machine would have acted as a random number generator (and not a cheap one). Lindley wrote:

> *Inelastic scattering produced, as expected, a complex mess of new particles,*

and Kendall, Friedman and Taylor showed that the statistical properties of this mess behaved in a relatively simple way at higher energies.

Much later, in 1990, they got the Nobel for it, which of course doesn't mean that anybody could learn anything from the outcome, since the scattering process is not understood. The calculations involving radiative corrections had been "difficult and tedious," as Taylor writes,[43] but for sure, they were missing one thing: correctness. A reliable theory of how accelerated charges radiate just does not exist. This kind of research is characteristic of the high-energy business: forget about the unsolved fundamental problems, do some high-tech experiments with the latest gadgetry, and try to justify the arising clutter with vague arguments sufficiently resistant to be disproved by concrete numbers.

> *He uses statistics as a drunken man uses lamp posts – for support rather than illumination.* – Andrew Lang, Scottish poet (1844-1912)

Divine Order or Moonshine Arrangement?

The huge number of particles produced in the 1950s and '60s started to make the saner people in particle physics feel uncomfortable, and one looked for patterns that could somehow revise the impression of an unreasonable turmoil. Because neutrons and protons transform into each other, they were declared to be one single particle that distinguishes in "isospin" (which contributed nothing to the understanding). Another empty word was invented for the fact that some particles just don't care about the distinction between weak and strong interaction. Some particles come into life with one, while departing it with the other, and just this should call into question whether the notions of "weak" and "strong" interaction make sense.[1] Physicists solved the problem, however, by introducing a new term, *strangeness*. At this point, physics had obviously given up logical rea-

---

[1] In plain words, their production is not that unlikely, but they live for quite a long time. You may ask, so what's the big deal? But for the contrived model, this was a problem.

soning and taken off to a metaphoric level, nicely wrapped in linguistic and mathematical jargon. Yuval Ne'eman and Murray Gell-Mann depicted the newly found particles in a diagram with the axes *isospin* and *strangeness*, whereby sometimes a pattern of eight lattice points showed up that in the following was named "The Eightfold Way" – it is not only the terminology that is close to esoteric here. Why squeeze two quantities that have nothing to do with each other into one diagram?

But back to experiments. Inelastic scattering had shown some unexpected regularity, called "scaling". Although scaling appears to be the consequence of a rather banal fact regarding physical dimensions,[44] the theoretician James Björken came up with a theory that claimed neutron and proton to be composite particles, and the phenomenon of scaling to be an essential prediction of it. Because his calculations were quite impenetrable, almost nobody listened to him, but this suddenly changed when the famous Richard Feynman, in a similar vein, considered so called "partons" as composite particles of the proton. The experimental fact here was still nothing more than that protons under fire felt much "harder" than expected. If one naively assumes the proton to be a smeared-out charge distribution – a kind of softball – the projectile electrons would not have been deflected as much as they were seen to. A nice trick, by the way: start from a naïve, unjustified belief, such as that electrons are hard and protons soft, and if your assumption turns out to be faulty, then deduce anything you want. In this case, it was concluded that the mass of the proton was not distributed homogeneously, but concentrated in several hard "scattering centers."

However, this is just a vague hypothesis. Remember that the very assumption of homogeneously distributed charge was untenable in view of what had been learned from quantum mechanics. The absurdity is often covered by claiming the situation is similar to Rutherford's famous scattering experiments, when he shot helium nuclei at gold atoms: "It is as if you fired a 15-inch shell at a piece of tissue paper and it came back and hit you." Rutherford correctly deduced that the gold atom's mass must be concentrated in a small volume of space. The comparison is flawed, however, in that the model assumed the proton to be a single amorphous cloud, quite different from the atom.[45] Additionally, the energy losses due to radiation of the accelerated charges are utterly unknown. Thus the result just boils down to the fact that we don't understand electrodynamics – one of the big insights revealed by postwar physics.

Democritus 2.0

Let us now consider a more general aspect. It was in the 1960s when the absurd enterprise began that tried to carry on the idea of elementary building blocks of Nature at a still smaller level – though the whole paradigm had long been destroyed by quantum mechanics. When should the division paranoia ever end? And the theorists' promise that this is the last round of splitting, from an epistemological point of view, is hard to take seriously. If you confront a diehard particle physicist with this question, he will weasel out by eagerly claiming that it is such a big insight that the notion of a "particle" has changed. Its properties, you are told, must be redefined to be compatible with the incredible theoretical progress made. Not the materialistic blocks of Democritus, no, the true nature of a particle would be a *collection of properties*. Just imagine particles as little packages labeled with tiny stickers; "hypercharge," "lepton number," "strangeness," "bottomness," pasted over the basic entities like charge and mass. What nonsense. When physicists call such constructs "particles" they just sooth their nerves by psychologically displacing how much they have departed from reason. Be it due to the huge number of particles, or be it due to the orchid-like terms needed to classify them, theory has long since left the realm of calculable simplicity that characterizes credible science. Describing has long since taken over understanding, but for doing true science that seeks understanding, an entirely different kind of researcher is necessary. Dirac, after reflecting about the proton and the electron, said at a conference in 1930:[46]

It has always been the dream of philosophers to construct matter from a single fundamental particle, thus it is not entirely satisfactory that we have two in our theory.

Two (!) particles were already too many for him. What would he have said about today's standard model?

Physics Goes Loopy

> ...*the descending size scale atom, nucleus, quark. The nasty suspicion arises that the thing does not end here ...* — Emilio Segrè

So how did it come about that today we are told that quarks are the elementary building blocks of matter? What followed is a textbook example of a symbiosis of vague experimental facts and theoretical fantasies with poor predictive power. The wishy-washy regularities that had been christened as "the eightfold way" were classified using the mathematical tool of group theory, or "symmetry groups," e.g. a description of how you may rotate objects in three dimensions (which would be called SO(3)). The math-physics symbiosis here drastically differs from other successful fields like, say, general relativity. There, you compare the predictions with the measurements and get a number expressed in percentages. Group theory instead just plays with qualitative properties of particles, a juggling on a noncommittal meta-level extraneous to physical mechanisms.

This was possible because theorists have abusively generalized the idea of symmetry groups beyond the realm where they made sense. It is perfectly ok to say that the laws of mechanics don't depend on the rotations in three-dimensional space, hence they are "symmetric under SO(3)" (the idea originated from the famous mathematician Emmy Noether). But it becomes ridiculous if instead of real space you apply it to the fancy diagram drawn with "isospin" and "strangeness" as axes. These symmetry operations[1] have become the dominating paradigm in theoretical physics, although they turned measurable predictions to completely metaphorical concepts. Richard

---

[1] The transformation of these particles in the pattern into each other is considered as "rotation" in that abstract space. Beyond the formalism, there is no underlying idea.

Feynman once said, about an alleged unification scheme based on such group theoretical concepts: [47]

> *SU(3) × SU(2) × U(1)? Where does it go together? Only if you add stuff that we don't know. There isn't any theory today that has SU(3) × SU(2) × U(1) – whatever the hell that is – that we know is right, that has any experimental check. Now, these guys are all trying to put this together. They're trying to. But they haven't. Ok?*

It is always a pleasure for me to quote that, when debating with particle physicists the alleged "stringent simplicity" of their model. However, much earlier than Feynman, Wolfgang Pauli had hit the point. He called the spreading nonsense "group pestilence." Theoretical physics has suffered for half a century from the infection.

As a consequence of the above ordering scheme, one may assume hypothetical components of the nucleus, which were called "quarks" by Murray Gell-Mann, one of the most aggressive promoters of the theoretical fancy (who was never bothered by the slightest doubts about his outstanding intellect). The "quarks" were then readily identified with the "scattering centers" of protons.

Nobody at the time seems to have reflected upon the grave epistemological defeat coming along with another, complicating subdivision of elementary particles. The idea of quarks does not explain anything, let alone provide a revolutionary perspective. It is precisely such fake understanding that, without being testable by a concrete observation, has eroded physics, the gradual spreading of the sickness being justified by the argument "we don't have anything better." Sadly, Richard Feynman, with a mind of refreshing criticality elsewhere, also capitulated, complying with the zeitgeist. "There is much evidence in favor of the existence of quarks and little evidence against, thus let's assume they exist."[48] As if you had to prove that a theory is nonsense instead of proving it makes sense!

> *A conclusion is the place where you got tired thinking* – Martin H. Fischer

The Higgs Fake

A Parroter's Guide to Writing History

*A part of the celebrities' glory is always due to the stupidity of the admirers. –
Georg Christoph Lichtenberg, German physicist (1742-1799)*

An ironic aspect here is that Gell-Mann's quarks, which had to be bound inside the proton, completely contradicted Feynman's original idea of partons, which he thought of as free particles in a (due to relativity) pancake-like deformed proton (a naïve assumption anyway). Once Feynman's authority had been widely used to publicize the concept, nobody cared any more about the inherent contradiction.[49] Instead, by ritualized parroting, the henceforth ineradicable self-deception of the quark model being a "simplification" gained ground among particle physicists.

Another thing much liked by the human psyche is retrospectively making up a reassuring narrative of what has often been a confusing and irritating story. For example, a big fuss was made about Gell-Mann's alleged "prediction" of the omega particle with an energy of 1690 megaelectron volts. Genuine mass predictions don't exist in all particle physics, and in this case, just a range was given that could be guessed as the continuation of a previously found chain like the position of the next lamppost in the street.

A theoretical development that contributed considerably to the acceptance of the new picture was the proof of the existence of "renormalizable theories" by Gerardus 't Hooft and Martin Veltman (Nobel Prize 1999). Many had racked their brains about this before, however, renormalization itself, established some 30 years earlier, is a fundamentally flawed concept. Thus however brilliant the math may be, there is probably no physics behind.

Whenever a generation of physicists dies out, a new layer of fancy may be established on top of something that had been considered absurd before. One would like to hear Einstein or Dirac's opinion about the contemporary ideas, but they are not just unfashionable, sadly they are dead. A "new class of renormalizable theories" would have them spinning in their graves.

Rather than predictions, the history of particle physics was full of unexpected problems. Just one of them was the size of the proton,[1] which was inconsistent with the data of the Stanford Linear Collider (SLAC). As always in such cases, an ad-hoc fix by means of auxiliary assumptions was invented in the form of "gluons" (literally glueing the loose ends of the theory) and quark-antiquark pairs, so called "sea quarks." At that time, nobody cared any more about the increasing abundance of new particles, as Andrew Pickering noted[50]:

> By this stage, the quark-parton model was in danger to becoming more elaborate than the data which it was intended to explain. A critic could easily assert that the sea quark and gluon components were simply **ad hoc** devices, designed to reconcile the expected properties of quarks with experimental findings. Field theorists could argue that the sea and glue would be required in any sensible field theory of quarks (although they had no actual candidate for such a theory...).

Wannabe Makers of the Laws of Nature

Things had come to a pretty pass at this stage. The reader of today's textbooks of particle physics has no idea how many theoretical concepts, each of them more arbitrary than the previous one, have been discussed in the history of the field, such as "Regge theory," "S-matrix approach," "bootstrap theory" etc. All this stuff lies in the graveyard of obsolete theories, though physicists never write explicit obituaries in journals. And it is often just coincidence which theory is chosen to mitigate a contradiction.

One idea that survived and flourished was "weak neutral current," the assumption that neutrinos may transmit their energy to neutrons as well (well, what's the big deal you may ask justifiably, but...). Though an entirely qualitative idea, the neutral currents were highly desired by theorists who liked to support their abstract playing with mathematical symmetry groups, in order to establish a theory of "electroweak unification." The term is both one of the most parroted and the most misleading in physics. There is no such unification,

---

[1] The value is still disputed today, see R. Pohl et al., Nature 466 (2010), p. 213.

The Higgs Fake

unless you greatly hype a teeny-weeny, hardly quantifiable effect with questionable methods. One of the most profound methodological flaws of particle physics is the focus on arbitrarily chosen, fashionable phenomena. Not the mass or charge, no, it is some utterly irrelevant "R," "y," or "omega" factor declared as interesting by the celebrities, usually some relation of combination of enhancement of fraction of ... qualitative shit.

Because theorists all over the world really started screaming for neutral currents (Weinberg, Glashow and Salam later got the Nobel for it), an enormous machinery was set in motion, resulting in the construction of Gargamelle, a giant bubble chamber that was able to detect neutral particles with an unprecedented sensitivity. Indeed, a few pictures suggested that a neutrino created a shower of heavy particles. However, as Andrew Pickering points out in detail,[51] the shower of particles could in principle be caused by neutrons as well – in both cases, it apparently comes out of nothing. The facts were not as unambiguous as desired.

If I cannot say a priori what elementary propositions there are, then the attempt to do so must lead to obvious nonsense. – *Ludwig Wittgenstein*

The Neutral Sickness

Let us pause for a moment to elucidate a very general methodological problem of particle physics of which the neutral currents offer a striking example. Neutral particles, due to their incapacity to separate electrical charges from others, are always "seen" by indirect evidence. Whether a signal originated from a neutral pion, or kaon, from a neutron or some of the neutrinos, is decided by the theoretical assumptions. The more neutral particles a theory has at hand – a whole bunch by now – the more arbitrary the interpretation. And this is where technology makes its unhealthy contribution to the game. In such cases, where one could just honestly admit that there is no way to separate the different origins ("channels") of particle production, it is tempting to use complex computer models and make them do the undoable (a practice that continues to this day). In the simulations carried out for the analysis of the Gargamelle experiment, a series of unjustified assumptions entered the model, such as an energy cut-off at one gigaelectron volts and other estimates of physical parameters that had nowhere been measured.[52] After much effort, data analyzers came to agree that from 290,000 (!) photographs taken, 100 events contained the sought-after neutral currents. This is perhaps the most striking example of a "discovery" entirely based on the adaption of interpretative procedures[I] – an agreement in the community about a theoretical model. It was then, if not before, when the boundaries between theory and experiment ultimately blurred. Favored by the absence of testable numbers, there is no clear distinction any more between theory and experiment, and it is highly annoying that most scientists (and in turn, the public) still gullibly follow the idea that a certain "theoretical model" in particle physics is "experimentally tested." It has long since been transformed into a messy hodgepodge.

All testing, all confirmation and disconfirmation of a hypothesis takes place already within a system. – Ludwig Wittgenstein

Factual Constraint to Agree

The neutral current story is not only a lesson in scientific methodology, but also in sociology. It is highly interesting to follow the detailed account given in Peter Galison's *How Experiments End*. After the discovery of neutral currents had eventually been published, the most-read paper in physics that had *never* been published officially, arrived at CERN in December 1973. A group of American researchers had succeeded in reducing the unwanted neutrons with a new method and concluded that there were no neutral currents. This was alarming news, CERN's scientific reputation was at stake, and the director made rallying calls to the researchers. Eventually, the competing group changed their setting again, and after many back and forths[II] the community agreed that neutral currents did exist, while

---

[I]   In fact, the existence of neutral currents had been excluded earlier (see Galison, p. 166).
[II]   Later mockingly called "alternating neutral currents" (Taubes, p. 3ff).

The Higgs Fake

the contradictory paper was never submitted to a journal. It is interesting to listen to one of the authors, D.B. Cline:

> *As the results began to emerge, we were being pressed harder and harder for some kind of decisive answer from people. It is very hard to communicate to you how [things were], when you are in the center of the stage at a time like that, particularly in high-energy physics where you do not quite have control over your destiny. You have to work with collaborators, with the lab, with the director, with the program committee, and with all the people who do the chores that allow the experiment to be done. You're be leaned on over and over again to produce, whether you're ready to produce or not.*

I think this quote gives ample reason to reflect the degradation of modern "big" science from the era of true discoveries (and Gargamelle is a worthwhile case to study). It is clear that the sociological aspects in big science experiments could no longer be ignored (to put it mildly). The post-war paradigm of practical, technology-based physics rather than deep reflections on nature led physics deeper and deeper into trouble.

A few general aspects of the period until the early 1970s may be summarized. The cavalier production of particles had eventually come to be perceived as a problem, but not tackled at the roots. Theoretical concepts such as symmetry groups, with their just qualitative predictions (e.g. that a particle should exist) offered a superficial, symptomatic treatment at best. What seems the greatest absurdity of that era is how the new concepts were christened "observations." Considering the many layers of arbitrary assumptions under which group theory connects to reality, one might as well say that they have nothing to do with each other. Establishing such purely metaphorical reasoning when comparing experiment to theory is probably the most severe sign of the latter's degradation. "We have made some progress since then," is the particle physicist's stereotypical answer. But a Newton, a Maxwell, an Einstein, a Dirac or a Bohr (let alone Pauli with his sharp tongue) would never have considered such stuff as science.

Alexander Unzicker

CHAPTER 9

APPROACHING FANTASY LAND

HALFWAY BETWEEN SCIENCE AND NON-SCIENCE

> *Concepts that have proven useful in ordering things easily achieve such an authority that we forget their mundane origin and accept them as unalterable facts. Scientific progress is often stalled for a long time by such errors.* – Albert Einstein

Despite the transition to metaphorical concepts outlined in the last chapter, reason was not yet eradicated completely from the particle physics community in the early 1970s. Some researchers still perceived the outlandish nature of the fashionable theoretical models. Remarks such as, "Who believed that parton light-cone crap anyway?" were still heard in discussions.[53] In 1974, that changed. As we have seen, so-called resonances, the increase of reaction probability (cross section) at a given energy, were usually interpreted as particles. A very pronounced resonance at 3.1 gigaelectron volts was discovered in November 1974, with a remarkable narrowness that indicated a long lifetime of the particle called J/psi. This event led to the general acceptance of both the quarks and their theoretical counterpart, quantum field theory, the essential part of a system which is now called the standard model of particle physics. Let us have a look how that happened.

A much discussed topic at the time was the so-called R-crisis, R being a number that indicates the ratio of the cross section (production probabilities) of muons and hadrons when electrons crashed into their antiparticles, the positrons. This is not a fundamental riddle of Nature, but rather one of many aspects of a collider experiment one could define as interesting – just an example of how academic data amalgamation, bare of any deeper reflection, creates its fashions. Theorists developed not less than 23 (!) models, predicting R in the range from 0.36 to infinity.[54] At the same time, everybody firmly believed it was a constant. The observational value of R ranged from 2 to 6, but showed (in particular data from the SPEAR collider) an annoying linear increase with energy that no theoretical model had predicted. This caused considerable irritation, as described by Andrew Pickering: "Electrons and positrons threatened to annihilate the quark-parton model as well as one another." Thus in the summer of 1974, particle physics was in a critical state. The discovery of the J/psi should resolve the problem. Why that?

The quark model until then consisted of three exemplars: "up," "down" (forming the proton and neutron) and "strange." The up and down were considered partners, and thus a companion for the strange was theoretically most welcome. Remember then that the R data were an anomaly, and every anomaly in scientific data – back to the medieval epicycle model – can be accommodated by additional degrees of freedom. Thus the expectation in the theoretical community was that a new quark, called "charm," could satisfy the necessities, long before the experiment came into the game. However, let's not repeat the common narrative; the link from the J/Psi observation to the c quark is basically a fairy tale. Since isolated quarks cannot exist (we'll come back to that absurdity later), all that one could expect is a "charm-anticharm pair" ("charmonium"). This is a typical Fata Morgana particle physicists use to fool themselves with. The appearance of a particle *together with its antiparticle* says little – for the very reason that any peculiarity of a desired particle is compensated by the respective anti-peculiarity of its antimatter twin, yielding an excellent excuse for any banal signal to be interpreted as *consistent* with the peculiar properties. For example, any particle pair decays to two photons, and thus if you postulate your great unified theory fiddle-fuck particle, its anti-twin boils it down to two nice photons. That's how it works to this day. The J/psi is therefore, remember, bogus evidence for quarks.

The Independency Tale

The legend recounts that the extremely narrow resonance at 3.1 gigaelectron volts was discovered *independently* at two different laboratories, thus enhancing the credibility of the analysis. I wonder why physicists don't react more skeptically to such Santa Claus coincidences, but here is the story, told by a researcher of a Munich Max Planck Institute, and recounted in Taubes's book.

Samuel Ting, with his group at Brookhaven National Laboratory, was the first to see the signal and planned to perform additional tests. Ting was a successful experimentalist, but probably the most disliked person in particle physics, and probably for good reason. Just to give an example, Ting contacted authorities in Beijing and demanded they withdraw the visa of a compatriot physicist who had participated in CERN's Christmas revue, where he had portrayed Ting in a harmless sketch.[55] There are other stories I can't tell because I don't want legal trouble. Just ask any particle physicist for such – sometimes unbelievable – anecdotes. Given that Ting "motivated" his collaborators in such a way, no wonder that little incidents happened. When Ting visited Stanford in November 1974, a member of the collaboration passed a small slip to Burton Richter's group, indicating an energy range, adding: "Look there, I've never been in your lab." Richter, who had the more powerful machine at SLAC, but no idea where to look, found the particle within some days, forcing Ting to a joint press conference where the discovery was announced. Apart from the satisfaction of receiving the 1976 Nobel Prize, Ting probably resented that he had to share it.

Celebrated Artifacts

As is already clear from the prompt reaction of the Nobel Committee, the event was followed by incredible hype. The fact that this additional quark was another complication of the model did not even enter the theorists' minds at the time. The predictions regarding the "charmonium" were plain wrong – the peak in the diagram was 40 times narrower than expected,[56] and the excitation states were blatantly aside from the calculated positions. But their sheer existence was celebrated as a triumph of the model.[57] The DORIS accelerator in Hamburg detected an accompanying particle, the η_c, which was seen as another confirmation of the big discovery. (The mass was different from the prediction, which was seen as an interesting feature rather than a problem.) However, data from SPEAR in 1979 demonstrated that the excitations found in Hamburg were all artifacts, thus key evidence for the charmonium model was retrospectively destroyed.[58] But when the party is over and the Nobel has been handed out, who the hell cares?

It was also a perfect time for windbags surfing on the new fashion to make their careers. As tends to happen during revolutions, the loudmouths get the leading positions (Pickering gives some examples). The guys who sit and wait for the latest hype, and then immediately start traveling from CERN to Harvard and back to give their enthusiastic talks – aren´t they the *exact opposite* of mavericks like Faraday, Newton and Kepler, who hesitated to present their findings after years of isolated, modest study of their subject? Physics has become so sick.

Seasonal Predictions

Let's see how theory contributed to the "November revolution." In 1970, Sheldon Lee Glashow, John Iliopoulos and Luciano Maiani had proposed the GIM mechanism (after their surnames) which required a fourth quark. As always, the underlying reason was a contradictory observation (there was a problem with strangeness), and as before, people carelessly disregarded that free parameters make the theory less credible (It seems to me a similar mental deficiency that allows central bankers to continuously print money, while being convinced that it's ok for the economy). And as always, their "predictions" had been wishy-washy to the point that nothing could prove them wrong, once their particle was identified with that something discovered by Ting and Richter. Particle physicists have the predictive power of meteorologists who see summer and winter coming. And they have a great deal of flexibility in accommodating results. The wannabe makers of the laws of Nature such as Steven Weinberg have always invited tinkering with the model whenever a contradiction popped up.[59]

The absurdities don't end here. Guess what followed? A fifth quark, "naturally" a member of a "third family" of quarks. Why that? In the meantime, another species of particle hunters, specialized in chasing a different kind of particles called leptons (light particles), had concluded that their data required a third particle *tau*, besides the existing electrons and muons. As usual, they hadn't seen anything except that some energy was missing, and this missing something (64 events of it) was declared to be the new particle. It took group leader Martin Perl a while to convince his own group of the interpretation. The argument was essentially declaring one interpretation as "very natural," and an alternative one which denied the existence, as "contrived."[60] In 1976, several groups were looking for the tau at the DORIS accelerator, and found nothing. Luckily, soon afterwards SPEAR interpreted its signals as taus, and the community agreed on their existence. As always, the pendulum swung towards the more comfortable solution that something was real rather than an artifact. For people who want to accuse me of promoting a conspiracy theory: I don't deny some measurable physical phenomenon related to such a "particle." But continuously producing new names instead of daring to think that the whole system is crap is nothing but psychological repression. It just makes no sense to call it a particle – as it did not make sense to call a poorly understood ellipse an epicycle.

A Growing Family

With the tau (a heavy electron) a member of a "third family" had been born, stimulating the greed of the quark hunters to look for such a species, too. Emilio Segrè, the discoverer of the antiproton and one of the very few reflective (and therefore unheard) minds in the community, wondered:[61]

Why should Nature produce two different particles which are distinguished in mass only and are identical otherwise? These are all examples of unexplained and unconnected facts. Some time ago, a third lepton was discovered ... is this the beginning of a finite or infinite chain of leptons?

With the establishment of the third "family" of light particles it was clear that the hunt was on for the fifth quark. The saga of the J/psi had accustomed the community to the discovery of new massive hadrons. In 1976, Leon Lederman and his group from Fermilab identified a "Y particle" – supposedly a pair of the new "bottom-antibottom" quarks – at an energy of 6.0 gigaelectron volts. When the apparatus was restructured, the signal disappeared, but another cluster around 9.5 gigaelectron volts popped up, and was – guess what? – interpreted as the "Y particle." Why distrust a group that had shown experience in detecting particles? The discovery is now regarded as having taken place in 1977, though an embarrassing story followed. Another hump at 10.0 gigaelectron volts was interpreted as "excited state" Y' (always a cheap explanation that does not do much harm), with the consequence that another "excitation" Y" should be visible at 10.4. Unfortunately, that region was just out of the range of the DORIS machine, while its successor, designed for energies up to 40 gigaelectron volts, was not sufficiently sensitive to detect such a relatively low mass – according to the post-hoc explanation of the failure. Soon afterwards, however, the Y, Y', Y" and Y''' (10.55 GeV) were found at the CESAR storage ring in Cornell. Why I am telling you all this? There might be people who find that back and forth convincing, but I am not among them. It is clear that the positive, confirmatory results were always given more credit than the negative ones, and it is annoying that in most cases, the reasons for the latter remain not only unknown but also uninvestigated. Given the considerable number of artifacts, it is quite likely that the Y-bottom quark story was just a wannabe copy of the J/psi-charm quark saga (there was also a Nobel that Lederman got in 1988). I cannot prove my suspicion, however (who can?) against the angry assurances of experimentalists – let's assume for a moment they are right, and that everything is clean and not an artifact. But forcibly interpreting the phenomenology as another quark is still stupid. At best, it just means that there is a process related to the elementary particles (and I mean the real colliding ones, electrons and protons) we do not understand. It is clear that postulating another complicated ingredient doesn't make sense. If the progress was not incredibly slow at the same time, researchers would probably become aware that they are developing another epicycle model. Here again, the huge size of their machines, which take so long to build, protects high energy physicists from having to think.

Toy Bricks and Cocksure Glory

If you are by now hoping that the absurd complication has reached its peak with the third quark family, you are going to be disappointed. It is Wolfgang Pauli's fault. Not that he liked complication (the contrary), but his exclusion principle which won him the Nobel Prize in 1945 says that particles with a spin $1/2$ (quarks have half-integer spin) could never be in the same quantum state. Some handicraft particles of the model makers, alas, were in contradiction with that principle.

How do theoreticians tackle such a problem? They simply imagine that all quarks come in three "colors," red-green-blue, which are, of course, unobservable. Yes, they have tripled the number of quarks by doing so, but this kind of qualm doesn't cross their minds any more. The auxiliary "gluons," therefore, must be imagined as colored like red-antiblue, and the whole menu is decorated by some ad-hoc rules such as that there are no single-color gluons, and the postulate that all three colors must show up in a neutron or proton. They pull an arbitrary rule out of their ass, just to furnish it with a still more arbitrary restriction right after. Theoretical particle physics is full of such special laws with special exceptions, symmetries that are asymmetric because they are "broken", featured non-features, dry water and pregnant virgins – a jungle of dark logic in which they have lost their orientation.

The most bizarre thing of all is to call a particle something that is not a part of anything. You cannot divide a neutron into its quarks. This is not linguistic nitpicking, but there should be a mechanism, a physical reason, why the hell single quarks don't show up. There isn't any. In the worst tradition of bad philosophers, the absurdity is wrapped in a vacuous term, *confinement*, and that's it. Quarks cannot leave the nucleus *because* they are confined, a textbook example of a doctrine, but it is nevertheless accepted by our brightest minds, the maverick physicists. David Lindley mocked the concept:[62]

> *In the end, the quark model succeeded by the ironical trick of proving that no quark would ever be directly seen by a physicist. This liberated physicists from any need to demonstrate the existence of quarks in the traditional way.*

The Wikipedia entry appears to be a satire: "The reasons for quark confinement are somewhat complicated; no analytic proof exists that quantum chromodynamics should be confining." The fact that it became the established theory clearly marked a time when the skeptical intelligence still present in the 1970s had passed away, and a streamlined generation of non-thinkers took over (somewhat like in the 1930s, but worse).

Even if it is just one thing quantum chromodynamics cannot explain, let's be clear: there is no reason for confinement whatsoever, even if David Gross, Nobel laureate of 2004, likes to suggest that "asymptotic freedom," a theoretical fashion he has founded, has something to do with it. In a tour de force of brilliant math Gross was able to demonstrate that there could be forces that become infinitely small at zero distance – while it would be of real interest to understand why the quark forces become infinitely great at finite distances. It is as if you are asked to explain the height of Mount Everest and you deliver a proof that the sea level must be zero. Theoretical physics has reached new horizons.

Alexander Unzicker

Unwanted Experimental Meddling

The whole story took quite an interesting turn in the late 1970s, which is almost forgotten today. Scientists had searched for isolated quarks — and found them! According to theory, quarks should bear fractions of the elementary electric charges e like 1/3 e, and a group from Stanford built a highly sophisticated version of Millikan's legendary experiment of 1920. A competing group in Genoa however found no such hints, and it is highly interesting how Andrew Pickering describes the scientific dispute.[63] It is a cruel lesson for every naïve realist who might believe that an experiment always delivers unambiguous answers, rather than being biased towards the researcher's expectations. Ultimately the fractional charges did not convince the majority of physicists, but the sad aspect here is that the highly respectable effort was casting pearls before swine. Theoreticians like Gell-Mann didn't want to be bothered anymore with their confinement ideology. The experimenters had become a nuisance, like an honest finder who returns a wallet to the pickpocket who had thrown it away after having picked it clean.

Murray Gell-Mann, shrewdly, had assured himself of both outcomes by encouraging physicists to look for isolated quarks, in order to "demonstrate their non-separability." Savor that! In any case, he was to turn out to be either the prophet of confinement or the creator-predictor of the new particles, a glorious role he surely saw himself in. The cover of his book, *The Quark and the Jaguar*, calls him the "Einstein of the second half of the twentieth century." Gell-Mann wasn't beset by self-doubt when he swaggered about Einstein's failing abilities when he worked on a unified field theory. Einstein was in fact at that time twenty years younger than Gell-Mann was when he wrote about string theory. Gell-Mann's mental capabilities by then just sufficed to rehash that multidimensional nonsense. Well, that half-century is over, fortunately.

The Higgs Fake

High Priests in Action

Gell-Mann, after his quark proposal derived from symmetry group considerations, had continued pioneering the metaphorization of physics by linking non-quantifiable concepts to vague experimental predictions. He himself gave a telling account of that bizarre reasoning:[64]

... we construct a mathematical theory of the strongly interacting particles, which may or may not have anything to do with reality, find suitable algebraic relations that hold in the model, postulate their validity, and then throw away the model. We compare this process to a method sometimes employed in the French cuisine: a piece of pheasant meat is cooked between two slices of veal, which are then discarded.

It seems to me that some theorists believe that coquetting with insanities justifies their application. There was a time, probably, Dirac and Einstein's, when the search for the laws of Nature was a serious responsibility. Research is not a thing for brilliant idiots to play with. I am not upset in the first place because I believe these theories are bullshit. It's the attitude of today's researchers that upsets me: "Well, it could be bullshit, but anyway we theoretical physicists are the brightest bulbs, so what?"

In this chapter, we have reviewed the rise and establishment of the quark concept. Here are some key points: another unjustified subdivision of the elementary building blocks is absurd. The inconsistencies of electrodynamics, in particular the unknown radiation of accelerated charges, are swept under the rug. The classification by means of isospin and strangeness (entirely unrelated, besides being nonsensical quantities) leading to "the eightfold way," is esoteric. The colors are an inherently unobservable, arbitrary complication. The theory allows for an unlimited number of quark families, which is absurd. The short lifetimes make quarks (pairs) indistinguishable from instrumental artifacts. There are no quantitative predictions of the quark model whatsoever. The postulate of confinement is nothing other than a dogma. If you count all this – I think these are eight good reasons to consider – the quark model is eightfold crap.

Alexander Unzicker

CHAPTER 10

TOWARDS THE SUMMIT OF ABSURDITY

ARBITRARY THEORIES AND BANAL FACTS

Though this be madness, yet there is method in't. – *William Shakespeare*

After quarks had definitely been established with the "November revolution" in 1974, high energy physics needed a new goal to achieve. The bottom quark discovery in 1977, also due to its somehow embarrassing circumstances and the fact that it seemed too much of a repetition of the 1974 event, was quite unspectacular. Another feature of the standard model, in particular of Weinberg's theory of weak interactions, was the existence of extremely short-lived particles, the so-called intermediate vector bosons W and Z. The story of the establishment of the W and Z particles, rewarded with the Nobel Prize in 1984, is told in detail in the book *Nobel Dreams* by Gary Taubes. We shall have a look at it because, first, today's results (including the Higgs) rest on the physical reality of the W and Z. Then, I believe the story allows a couple of inferences of how physics is done today at CERN. Given that the number of scientists involved has risen from several hundred to several thousand since then, and the experiments are harder and harder to oversee, the situation described by Taubes can hardly have much improved.

One of the leading experimenters back then, Pierre Darriulat, complained that the world had lost its critical judgment about what was going on in the collider experiments:[65] "Nobody wanted to know the details of the experiment, they only wanted the results. You can say almost anything, and you don't feel among the physics community a strong desire to understand." People were dreaming of so-called "gold-plate events" – perfect, clean discoveries. Sadly, the experiments rarely satisfy such a simple expectation. As Darriulat added, the attitude of seeking results without explanations was "worst at the CERN management."

The Brute Force Hunter

Ambition is the death of thought. – Ludwig Wittgenstein

The central figure of the search for the W and Z particles was the Italian physicist Carlo Rubbia. Born in 1934, when the two-particle world had just ended, he received his PhD in 1959, when the number of particles was exploding. There is no sign that Rubbia had done any reflection on that, rather he seemed to be ardently engaged in the fashions of the time. And he didn't feel qualms about becoming famous at any rate. Several researchers have confirmed that he invented data for seminar presentations in the early 1960s, an issue that did not perturb Georges Charpak, the inventor of a modern detector: "High energy physics is full of stories like that."[66]

What is sure is that Rubbia was quick in apprehension, energetic and extremely ambitious, a workaholic. He understood pretty early what he later expressed with stunning frankness: "Since the times of Lawrence, if you want to be successful in our business, you had to have the biggest and most powerful accelerator." In 1960, Rubbia went to CERN. One of his undisputable merits is to have abolished an utterly misguided method of dividing accelerator experiments that was practiced at the time at CERN. Already back then, many research groups all over the world participated in analyzing the results, and since from the collision point particles fly apart in all directions, proper justice in distributing that data was done by assigning to each group a solid angle in space where they could observe the outgoing particles – it is like if 1000 art enthusiasts watching a Kandinsky painting were allowed to look at one square inch each. Rubbia brought CERN to use the so-called 4 pi-detectors (covering all spatial directions), which were standard in the US labs. Now he had the best machine in the world, and his goal was in sight.

> *Rubbia could yell louder than any of his colleagues at CERN, and usually think faster, and the result was that he rarely lost an argument.* – Gary Taubes

> *His numbers are what they are. They are usually wrong.* – *but if they suit his purpose, nothing is wrong* – Bernard Sadoulet, a colleague of Rubbia[67]

Rushing to Measure the Missing

The characteristic of the W boson was that it should decay into an electron and (an invisible) neutrino after 10^{-25} seconds. Though such absurdly short lifetimes had been accepted earlier, for the first time a huge project was undertaken in search of such particles. Keep in mind that W bosons (like Z bosons) can never get into a detector, but they are supposed to have a "signature" of an electron that emits an (invisible!) neutrino. Go figure. Thus all you need is the simplest particle in the world coming out of the collision, and something that is missing. This seeing by not-seeing is one of the most absurd developments of scientific methodology.

Needless to say, such an effect based on inferences was designed for misinterpretation in the kind of hasty analysis Rubbia was known for. What astonishes me is that, despite harsh criticism of Rubbia's methods (and style and demeanor, but ok), nobody addressed the basic problem of the misdirectness of such an identification. It seems that high energy physics already had enough absurdities.

Meanwhile, Rubbia defined how the identification had been done. To put it mildly, he didn't greatly appreciate dissenting opinions[68] – it is attested that he threatened to throw out people who weren't obedient to him. It wasn't easy to understand the outcome, though the UA1 central detector was a marvel of technology (as are ATLAS and CMS today). Taubes describes the situation:[69]

Physicists ogled the pictures that came out of it. But they didn't necessarily understand what it was telling them. They took Rubbia's word for it. And Rubbia was using his nose, saying, "It smelled like a W," or, "I feel in my bones that it's a Z."

In late 1982, the machine was about to deliver its first results. Rubbia urged his collaborators to work day and night before his visit to various institutions in the USA. He took a picture of a "W-event" with him. There, Steven Weinberg, Abdus Salam and Sheldon Glashow all happily agreed that it was the long sought-after W boson (which confirmed their theory, by the way), and a physicist of the US Energy Department was also impressed (high energy physics always needs to keep a close contact with the funding institutions). Burton Richter (1976 Nobel laureate) instead wrote on the picture bluntly:[70] "It's not a W. Burt." If one sees all this expertise in practice, it seems naïve to assume that physicists have a more reliable judgment than art critics who can identify a Rembrandt painting from its brushstrokes. Later it became clear that Richter was right. That didn't keep Rubbia from including the picture, knowing it to be wrong, in his collection of W-events to strengthen the statistics.

At that time, computer programs began to implement the – refined – criteria whether neutrino energy was missing and the thing had to be declared as a W event. The practice of detection by non-detection had been fully established and automatized. After all, the W had to be there and had to be found sooner or later. Remember that no W would mean that the theory of electroweak interactions for which the Nobel prize of 1979 was awarded was wrong. Inconceivable.

Rigged Reality

Another intense round of data interpretation was performed before the conference in Rome in January 1983. Rubbia's reputation was miserable in high energy physics, due to the fact that several times he had boasted prematurely of results that turned out to be false. His talk was met with overt skepticism in Rome. Leon Lederman commented:[71]

The speed with which the data was analyzed and physics presented was truly astonishing, considering the complexity of the collisions, the sophistication of detectors, and the hordes of experimental physicists.

It was obvious that Rubbia cared much more for being the first in the race than for reliable data analysis. In view of the competition between Rubbia's group UA1 and the (more cautious) detector group UA2, Lederman mocked the situation with a parable about two physicists confronted with a fierce bear. "Let's run!" said one, and the other, somewhat pedantically, responded, "But you can't run faster than a bear!" To which the first physicist responded, "I don't have to run faster than the bear. I have to run faster than you." It turned out to be a very apt allegory.

After the conference, the official announcement was to be given in a common seminar of the groups UA1 and UA2 (does that remind you somehow of July 4, 2012?). Rubbia presented the evidence for the W. Tellingly, he had been reassured beforehand by the director general of CERN, Herwig Schopper, that UA2 would not have contradicted his results.[72]

Rubbia managed to convince the community in a stirring and enthusiastic talk. His work was done – except for getting rid of his rival. The next day, Rubbia showed up in the CERN cafeteria with a sorrowful face, telling everybody he wasn't that sure about the discovery and he would delay the publication in order not to endanger the group's reputation. Rubbia had already sent the paper for publication by courier mail in order to establish priority,[73] but the hoax helped to fog UA2. Di Lella and Darriulat, the leaders of the UA2 collaboration, who had cleaner and better data, were fooled. UA1 was faster because it had no standards. Rubbia defined them at will. Rubbia was a man of action, not reflection, preferring rushing over thinking, his goal was to detect quickly, not to check thoroughly, personal ambition counted much more than the wish to find out something, and when necessary, he cheated. I wouldn't even call him scientist. Char-

Mission Accomplished, Science Dead

acters like Faraday, Rutherford or Bethe would have been ashamed by such behavior.

The story was repeated, more or less, with the discovery of the Z particle. Di Lella tried to be clever, writing a discovery paper with just the blanks to be filled in with numbers once the experiment delivered them. UA2 also hacked the UA1 computers to get some privileged knowledge. But they were scooped again by Rubbia (and again, CERN management provided him with the UA2 results privately to avoid potential embarrassment).[74] He had won the Nobel race.

Given the amount of prestige, power, and money at stake in CERN's 2013 experiment, I just cannot believe the grandiloquent assertions that today everything is clean, independent, and follows the rules. Things won't work so differently. But it may be some time before we know the whole story.

With the help of the director general (who carefully limited the names showing up in press releases to two; Rubbia and van der Meer) providing handy publicity in 1983, it was clear that Rubbia would get the Nobel, barring some unforeseen catastrophe. The "danger" was that something else potentially important was discovered *not* by Rubbia first. A phenomenon called "monojets" created considerable hype (it turned out to be nonsense later, and was forgotten by the selective memory of history). Rubbia was afraid of coming second and published so hastily that the members of the collaboration didn't even have a chance to see the paper. The calibration, the heart of a sound analysis, was so superficial that Michael Spiro, a UA1 member, telexed the journal *Physics Letters* to get his name off the publication. A theorist from Fermilab in Chicago later said: "I won't talk about this paper. It's crap. And it is well known that it's been published just for the Nobel."[75] On October 17, 1984, the prize was awarded to Rubbia and van der Meer.

Alexander Unzicker

What Happened Thirty Years Later?

All this, at least by the standards of the discoveries at the beginning of the last century, is a scandal. It is not only that sloppy and superficial work, trickily managed and full of swagger, prevailed over careful, thorough analysis. But CERN, back then, was already a Nobel-greedy big science company seeking to get close to politics and big money. Scientifically, a sick monster. It had nothing to do with the search for the fundamental laws of Nature undertaken by Faraday, Newton, Maxwell, Einstein and Bohr. Given that at CERN in 2013 the very same external motivations are at work, for a historian of science there is not the slightest evidence that things have changed.

In a debate[76] with a CERN spokesperson and ATLAS member, I briefly touched on Rubbia's way of working. He answered that certainly not everybody liked Rubbia (and he added, Ting), but nevertheless he considered the W a great discovery, even if it had been quite shaky at the moment of detection. And, if it hadn't been such a great discovery, the Nobel wouldn't have been awarded for it. You can't argue seriously with these guys with their circular arguments. But I don't believe it either. Once a theoretically much desired particle is established, they would never back out and recall. As Taubes phrased it:[77]

> But nobody ever won a Nobel prize for proving that something didn't exist or by showing that something else was wrong. And if anyone knew that, it was Rubbia.

Let's Assume the Best Case

I must clarify an important point. Let's assume for a moment that everything went cleanly, Rubbia was honest, every detail of the experiment was reliably tested and cross-checked etc. Even then, all this is still nonsense. Declaring a signal that consists of just an electron and some missing energy a new particle is something you can continue doing forever and with any number of particles. Given unlimited energy, the lack of understanding necessarily creates unexplained

The Higgs Fake

signals waiting for new names to be assigned to them. That's how it works.

A major methodological degradation that took place during the W and Z search was the invention of the trigger. There were just too many events to be stored and analyzed in a reasonable time, thus a sophisticated electronics was developed that looked for the "interesting" events. There is no historical, methodological, logical or other justification for triggering, besides "we have more data than we can process." Just produce less! Triggering is another element that narrows down particle physicists' perspective of open-minded research, an excellent tool for fooling themselves by focusing on the selection of data they wanted.[1]

One last irony is worthy of mention here, since later we will look at the parallels during the search for the Higgs. In late 1984, the zeta particle (which later turned out to be an artifact) was discussed in terms of it being the Higgs or a Higgs (or some supersymmetric version of it).[78] Taubes comments:

> *No one knew yet whether the original signal had been a misunderstanding of their detector and its capabilities, whether it had been created subconsciously by the physicists, or whether it was something more complicated. But it was gone. That's all they knew.*

If you confirm something, hype it. If you miss something, just fuhgeddaboudit. That's the biased strategy we still see today. That particle physics was left to the regime of people like Rubbia was also due to the fact that the standard model builders, in particular Steven Weinberg, rather lost interest in their byzantine construction and dabbled in cosmology instead, infecting that field with notoriously untestable speculations such as inflation theory. Roger Penrose once commented: "Inflation is a fashion high-energy physicists visited on cosmology. Even aardvarks think their offspring are beautiful." Today, cosmology and particle physics exchange their free parameters

---

[1] As happened in the "positron lines" disaster at the GSI Darmstadt, a German particle accelerator („Vanishing lines", www.gsi.de/documents/DOC-2003-Jun-25-4.pdf.)

like viruses exchange their genes, becoming thus ever more resistant to falsification. This is another story, but when you hear particle physicists repeating fancy stories about what happened 10^{-35} seconds after the big bang, it provides additional evidence that these guys are ready to parrot any nonsense.

Topped Baloney

The story of the W and Z particles was repeated, *mutatis mutandis*, with the discovery of the top quark that was announced in 1995 by the groups D0 and CDF working at the Tevatron in Chicago. There was considerable back and forth when deciding whether they were dealing with a top quark or not. In 1994, one group had tried to discuss four papers written by different members, in order to keep the process open and transparent; the result was that everybody was contradicting each other, an event remembered as the "October massacre."[79] (No wonder particle physicists returned to the streamlined system.) Ultimately, the top quark had to exist because the bottom quark needed a partner, as the Ws and Zs had to exist because otherwise the standard model was wrong. The top quark had two characteristics. The first was its extraordinarily high energy of 173 (!) gigaelectron volts, outperforming by orders of magnitude the other quarks; up (0.3), down (0.3), strange (0.5), charm (1.5), and bottom (4.6). Not only were these numbers all unexplained, but their weird distribution should indicate to anyone with his or her head straight that the model was outlandish. But, since the standard model didn't predict any mass, physicists could just go to higher and higher energies until they found what they wanted.[1]

Secondly, the top quark was eventually seen to decay into a bottom quark and a W boson – precisely that teeny-weeny, questionable signal that had been established at CERN some ten years earlier. That's how the system works: you assign a signal to one particle, A; than you do another experiment at higher energy where you remove all the A particles as background. It turns out that this does not de-

---

[1] Particle physicists claim that there was a prediction, but this is basically a fairy tale.

scribe the outcome, thus one baptizes the remaining signal as particle B. The next experiment, at higher energy again, removes As and Bs and their combinations as background. Call the remainder C. Particle physics also used the Greek alphabet and mixed the order, but the system is essentially that.

Not only it does it seem that particle physicists can find evidence for any concept if they search long enough, but also there is no result of high energy physics that cannot be accommodated by a suitable extension of the existing models, that is, at will. For every opulent meal of data the colliders serve, there is a theoretical dressing on the menu, and nobody asks where the fatness of the standard model comes from. With the sixth quark, the edifice was complete, yet the futile game had to continue. The show had to go on. The biggest absurdity – although we are already used to this – is the minuscule lifetime of the top quark, taking it again into the realm of pure interpretation by means of its decay products. The lifetime is so small that the top quark cannot decay into the usual hadrons. This seems to be another conceptual flaw (not predicted by anybody!), but you can turn the thing around and make a virtue out of necessity. As a member of one of the discovering groups phrased it: "The sheer enormousness of the top's mass makes its decay a fertile ground for new particle searches."[80] They believe they can continue with their idiocy forever.

CHAPTER 11

THE HIGGS MASS HYSTERIA

IF ANYTHING, THE HYPE OF THE CENTURY

On July 4, 2012, at the famous CERN seminar, scientists applauded, cheered, celebrated. The news spread quickly all over the world that the Higgs had been discovered (nobody cared about the subtleties of "the Higgs" and "a Higgs"), allegedly the verification of an almost 50-year-old idea formulated by a Scottish theoretician. The nonsense starts right here. It's not that the physics world had desperately sought the Higgs for five decades. Feynman, for example, died in 1988, and was never heard to mention the Higgs.[81] Rather, after the top quark was discovered in 1995, something had to be found in the theoretical boxroom to inspire the next round of high energy experiments. And a nice thing to play with was the "Higgs mechanism," even though it was not exactly an ingenious idea. Peter Higgs appears to be a modest old gentleman who honestly wonders how all this hype has fallen into his lap, but he is certainly no Einstein. You cannot compare a life full of passionate struggling with the laws of Nature to one single idea which was in the air.

And of course, there is an irrelevant meta-story floating around about who might have published a similar or the same idea before or

The Higgs Fake

after Higgs: Brout, Englert, Guralnik, Hagen, Kibble and anyone else who wanted to add his name to this baloney in order to get rewarded. This just means the idea was quite obvious in the jargon of the day and many picked it up, prepared or reinvented it, like Nambu, Weinberg, Veltman, Gell-Mann (according to him) and others. My preferred abbreviation is BEGGMHHKN'tHVW. A favorite topic of all the blogger-blabbers was how the particle should be named and who deserved the Nobel Prize. As he had probably done several times earlier, Nobel started spinning in his grave again on October 8, 2013.

The Emperor Is Naked

The field theory cornerstones of modern particle physics are based on a completely metaphorical level, upon which you may develop any fantasy to explain your needs. The deep problem in this case is that the standard model has not the faintest clue how to explain mass – the most basic property in physics is a closed book to today's physicists. Well, they say, it has to arise from something, from some "ground state," like a bullet at the bottom of a well (you have to be brainwashed by quantum field theory to appreciate such an unintelligible metaphor). Sadly, so far this explains the mass to be zero. Now the great idea: the surface, instead of being just a paraboloid-shape well, has a little bump in the middle, pushing the rolling bullet away from its comfortable position at the center and forcing it to roll around in a circular valley. You may now ask: Is the bullet the particle? No. What the hell is the surface then? Space? No. It's just a picture, something physicists imagine to visualize that it *might* happen in a similar way, the bullet on the surface may be *some* state on *some* field, but there isn't a real correlate. The Higgs mechanism is neither intuitive nor obvious,[82] as David Lindley, one of the few sane authors to describe it, states, "it may be as well regarded as mathematical invention." The only justification is that you can do some calculations, the preferred mind-narcotic activity theoretical physicists like to displace their ignorance with. But the less people can grasp a concept, the more they praise it. The "Higgs mechanism" rather than the Higgs particle, we hear, is the true thing of which we are supposed to admire the discovery. All this is easy to parrot. But nobody can explain why this Higgs mechanism should have anything to do with reality.

The Higgs mechanism is part of a still more general concept called symmetry breaking (simply because the bullet rolls off-center). There was once a particle physicist, Yoichiro Nambu, who had a background in solid state physics (this is considered a broad education), and he noticed that molecules are forced to choose an arbitrary orientation when they are arranged in a lattice. This is an example of a symmetry breaking, and where it occurs, it is readily understood by kindergarten children. Nambu now conjectured that this may somehow somewhere, in some state of some kind of field, occur in particle physics too, intoxicating the community of theorists who hadn't a clue how to resolve the mass problem.

From now on, everything in physics that was poorly understood was "explained" by "symmetry breaking mechanisms." More matter in the universe than antimatter? Must be a symmetry breaking. The proton heavier than the electron? Well, the symmetry is broken. Go to a doctor and tell him that your right leg hurts, while the left one is ok. "The symmetry is broken." You would send him to hell. No other science can afford such malarkey. To award the Nobel Prize in 2008 to such a dumb and self-referencing idea was one of the events that helped me lose my scruples and write this book.

The Science Fiction Ritual

So, for what was the 2013 Nobel Prize awarded then? The Higgs "discovery" probably was the most publicized event in modern science with the least meat. I am not going to retell the story with the kind of irrelevant detail that was broadcast to the world, such as the kind of font used in the ATLAS PowerPoint presentations or similar nuts. Let's try to understand what really happened – if anybody could understand this. After the discovery of the top quark (I continue to say "discovery" following common language; though "declaration of existence" would be more appropriate) in 1995, particle physicists had to sustain another long period of waiting. The Tevatron at Chicago tried to catch the Higgs until the collider was shut down, but the only thing that could be achieved was to exclude some mass regions. CERN could concentrate on the region between 115 and 160 gigaelectron volts, and early in 2011, almost everything had been excluded except a narrow band up to 130 gigaelectron volts, where the search was most difficult due to the large backgrounds. At that time, many had lost confidence that the Higgs was ever going to be discovered. Then, in December 2011, the first rumors spread that something had been found around 125 GeV, prompting a string theorist to do his territorial marking of the Higgs being predicted (yes, after string theory had failed to predict anything for 30 years) to have 125 GeV. He was awarded a prize by the American Physical Society[83] for that (why does the establishment not stand aloof from such nonsense? Because they have skeletons in the closet!). Anyway, let's not forget either that the theoretical masterminds had also covered their ass in case of a non-discovery, which was often declared as "possibly even more interesting than the discovery."[1]

In the following months, a vast amount of data was produced and at the end of June 2012 the first news came out that they had found a signal "compatible" with the Higgs boson. Some got upset because the information had been leaked prematurely to the public, but probably because this just demonstrated that the analysis wasn't as independent and secret as they wanted the world to believe. It is ridiculous to claim that there is no flux of information between the competing groups, ATLAS and CMS. Thus it wasn't such a big surprise that spokespeople for both, Fabiola Giannotti and Joe Incandela, came out with the same numbers – a five-sigma statistical evidence, just enough for a "discovery", in the jargon. It means that of such results just one in one million is wrong on average – statistically. The history of particle physics has seen a lot of five-sigma results vanish into thin air, but always for other, non-statistical reasons. This just means that the statistical figure is irrelevant for estimating the probability of the result being right or not. A stupid ritual, yes, but one we are used to.

---

[1] Ironically, the theorist James Björken in 1977 expressed the same kind of intellectual flexibility with respect to the W boson (see Pickering, p. 365).

Born to Decay

The phenomenon that most contributed to the five-sigma evidence was an extra pair of photons. Go figure. *Every* particle-antiparticle pair of any sort decays to two photons, thus to read into it such a peculiar construction as the Higgs boson is already outlandish. The only justification is that, at least since the W boson 30 years ago, high energy physicists have grown used to the absurdity of assigning rare but trivial phenomena to sophisticated theoretical fantasies. It is true that finding ever more subtle phenomena becomes harder – but this is not scientific progress to be proud of, it is an advance of technology. The more sophisticated your filters are, the more exotic particles you may read into your results. What is really dumb is to believe that this could continue forever.

The extensive filtering of the photon background however cannot be done correctly, as we have already mentioned in chapter 2 – physics just doesn't know how accelerated charges radiate. (The more readily physicists respond "of course we know", the more ignorant they are. Go and read your damn textbooks![84]) It probably could be kept in mind better if that was more recent news. But since the unsolved problem is more than 100 years old, people just fade it out.

Like the W and Z bosons and the top quark, the Higgs faces the problem of a short lifetime. The former particles decay long before they can leave the collision point, and therefore "detecting them" means interpreting their decay products, based on a series of inferences layered on top of each other. Though the Higgs "officially" has a longer lifetime (about 10^{-22} seconds, a number grown out of the heads of theorists, not from experiment), this is not what you see from the data. The quite broad signal (see fig. 2) rather suggests, according to Heisenberg's uncertainty relation, the same 10^{-25} seconds. People argue that the insufficient energy resolution of the LHC, alas, does not permit us to see the 1000 times sharper peak. But why did you guys build a machine with such a poor energy resolution of 6 percent then? High energy physicists respond with indignation that this accuracy is excellent and the best *possible*, which is probably true from their point of view. But they don't understand

The Higgs Fake

that this very lack of precision is one of the major methodological weaknesses of their murky business. General relativity or quantum optics have advanced to an accuracy five or ten decimal places higher. These fields are just better physics.

The analysis also contains a lot of theorizing about which particles created during the crash of protons successively turn into Higgs particles and their products. And because the energies at the LHC are in an unprecedentedly high region, there is no way to know that in advance unless you do optimistic extrapolations. My favorite example is that recent measurements (by a high-precision quantum optics technique) revealed that the radius of the proton is considerably smaller (about 4 percent[85]) than was assumed previously, Does that bother the slightest an experiment that crashes *protons* into *protons*? No. They come up with the subterfuge that the reaction probabilities are well known, blabla... and the radius is irrelevant. They deny the impact of real properties like *size* because it does not fit in the world of *their* protons they make up in their minds – because in that wacky world of particle physics size does not play a role. Whoever wonders about such a splitting of consciousness is, in their view, mistaken. Physics and particle physics are different planets.

Muddy Water

Since the big hysteria in 2012, further news has been delivered to the public that the particle was "compatible" with a Higgs, or had become more "Higgs-like," or that CERN director Rolf-Dieter Heuer "personally" had dropped the "like" (who cares?).

Such sophisticated semiotics about the nature of the Higgs are usually based on the "decay channels" measurements. "Channel" means the respective amount of Higgs particles that decayed to two photons (the gamma-gamma channel) or to four electrons or muons (the "four-lepton channel"). To get your own impression, look at fig.1 below. Being benevolent, one may feel that the data do somehow cluster around the dotted line prediction 1, though the agreement is all but spectacular, and sometimes plainly outside the error bars.

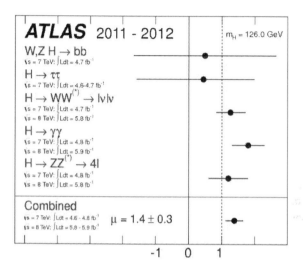

Fig. 1. Decay channel measurements of ATLAS (very similar to CMS).

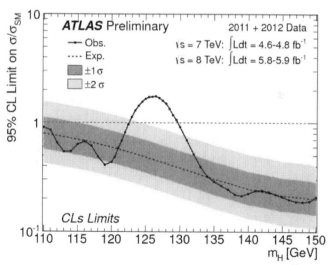

Fig. 2. Two-photon channel measurement of ATLAS.[86] The hump around 126 GeV should indicate the Higgs particle.

But in terms of the wishy-washy precision to which the field has been accustomed for decades, this partly contradicting data becomes "good agreement" or "nearly perfect consistency." But even if you are willing to accept such low standards, this is far from a clean procedure. The problem is that you can fiddle around with the decay channels, because they are not that easy to separate. For example, if the two energetic photons of the gamma-gamma channel later produce pairs of electrons or muons (which they can, but you don't know beforehand how many), the result would end up in the four-lepton channel. The "bottom quark channel" is merely a bunch of assumptions about the protons coming out of the proton-proton collision (what a surprise), and considered as the most "dirty" one even among particle physicists. Thus, in principle, the various channels may still shift their numbers to each other in order to comply with the observations. They are not really independent measures. Let alone if anybody had done a clean comparison of the number of independent measurements and the number of free parameters. The huge number of free parameters is a red rag for particle physicists, thus they prefer not to bother at all with such elementary questions of scientific methodology.

The Botch-up Mechanism

Let us come back to a more general aspect of the "Higgs mechanism," since it touches the very failure of the entire standard model. It is utterly nonsensical to claim that the Higgs mechanism tells us anything about the riddle of mass. (If somebody declares the Higgs as "the best idea we have about this" then he never had one.)

Einstein pondered for decades the equivalence of inertial and gravitational mass, one of the key problems to which the solution can only be found in a deeper understanding of gravity and cosmology. Particle physicists are not only ignorant about it, they are proud to be. It is an utter disrespect for the half of fundamental physics that is covered by hilarious boasting.

To the old enigmatic question why the proton is 1836 times heavier than the electron, we now get the enlightening answer that it's because it couples 1836 times more strongly to the Higgs field. Why don't people get mad when fed with such banalities? Probably because they live amid the madness. Or because they are hypnotized by some brilliant calculations. But calculations only make sense if you come out with a *number* – the Higgs stuff doesn't. It is unquantifiable, metaphorical junk.

In this context, there is another folly I cannot resist telling you about. Though neutrino physicists are now firm believers in their "neutrino oscillations," ironically, there is no place for the oscillations in the standard model. Despite all the theoretical jiggery-pokery, the Higgs mechanism cannot account for the masses that the elusive neutrinos are supposed to gain due to those oscillations. Thus physicists invoked another hanky-panky, called the "seesaw" mechanism. After I had expressed my doubts about such a botch in an email to the project leader of a huge neutrino-mass experiment (KATRIN), he wrote to me: "One of fathers of the seesaw mechanism is the Nobel laureate Steven Weinberg!" These authority-spoiled people think they are doing science when nodding their heads like donkeys to everything that has the smell of a Nobel perfume.

The Future of Particle Physics

> For the next step, still greater efforts will be required, without the guarantee that it will be the last step. The limit would be set by the cost, without the guarantee of ever reaching the end – Emilio Segrè,[1] 1980

But let us now be strictly scientific and make predictions. Here is what I would dare to forecast when particle physicists get approval for their new collider (I hope that governments will refrain from squandering further money, but who knows): There will be never clean experimental evidence for the theorized lifetime of $1.56 \cdot 10^{-22}$ seconds, which would require the hump in fig.2 to become at least 500 times narrower. What could happen, however, is that a similarly muddy signal like the Higgs is found *somewhere* and then readily de-

---

[1] See, as a contrast, Steven Weinberg's superficial arguments in www.nybooks.com/articles/archives/2012/may/10/crisis-big-science/

clared as a "second Higgs," "part of the Higgs multiplett," a "supersymmetric version of the Higgs" or whatever fancy the theoretical couturiers like to indulge. Peter Woit, a well known critic of string theory and SUSY, when poking fun at theorists who postulate higher and higher energies, is too naïve in believing that a nonsensical idea would never get its experimental evidence. (I must admit I also was too naïve when I took a bet against the Higgs in 2009). It is the very mechanism of forcefully assigning a fashionable meaning to data that, from a sober observational perspective, are banal signals. The floodgates are now open to let any nonsense pass. To summarize, if in 2017 there is another Nobel for a "SUSY" version of the Higgs, or some supersymmetric something, it is I who predicted it here.

The Dark Ages Are Still To Come

It may annoy some people, but the truth is that particle physics has come close to astrology. Exotic properties like "isospin" or "strangeness" define the existence of the particle as such, and all what matters if the classification within a pattern – in the same vein, the position of a new star may be assigned to a given constellation. The "eightfold way" of the quark model is not really that different from the twelve zodiac signs. You always find a name, be it the sixth toe of Ursa major or a louse in Hercules' coat. In the same manner as these signs cover the celestial sphere, for every particle energy there are various theories that allow us to digest the results with a bit of imagination, be it the fourth generation of quarks or a fifth force. In effect, every unexpected measurement of the last years has immediately generated "plausible" explanations, though statistics usually made them disappear like meteorites. The standard model is incapable of assigning a meaning to the mass of a particle, just as astrology has nothing to say about a star's luminosity. Particle physics avoids quantitative predictions because it would twist the knife in the wound of non-falsifiability, as it does with astrology. It is an arbitrary construction that will leave anybody frustrated who seeks insight. Maybe one cannot understand Nature, but for sure, particle physics cannot explain it.

Alexander Unzicker

PART III

ANTIDOTES

CHAPTER 12

WHO IS TELLING AND SELLING THE NONSENSE

THE BIG SHOW OF BOASTERS, PARROTERS AND FAKE EXPERTS

> *The difficulty lies not in the new ideas, but in escaping from the old ones, which ramify, for those brought up as most of us have been, into every corner of our minds.*[87] *– John Maynard Keynes*

The amount of propaganda distributed about the Higgs is overwhelming. Let's start with an almost amusing example. Michio Kaku, a famous string popularizer, said in an interview on CBS news that the Higgs particle was the "missing piece of creation" and it "put the bang in the big bang."[88] Whoever wonders about such claims should watch Kaku on Fox news, where he predicted the LHC would "show that our universe has collided with others where other laws of Nature held", to "tell us what happened before Genesis chapter 1:1" and so on. This is considered to be blatant nonsense even by particle physi-

The Higgs Fake

cists. Matt Strassler, a member of the CMS collaboration, complained about Kaku's "spectacular distortions" and wrote in his blog, "worse, Kaku presumably knew it was wrong." Well, scientists know, to say it with a quote of Barack Obama, who is the bullshitter. The problem for people like Strassler is that, at the end of the day, it's not easy to draw a line between Kaku and themselves. Kaku certainly represents the top level of loudmouth who does not even care to parrot reliably when making a spectacle of himself. But on the other side of the spectrum, the person who oversees both experiment and physics, the person who might explain thoroughly what happened at the LHC, just doesn't exist. It's too complicated.

Matt Strassler, in his blog, serves the public with seemingly detailed information on how the whole experiment works, but once you try to point to a specific problem (like the unsolved issue of radiating charges), he also declares that things are complicated, unfortunately. The idea that the entire model is on sandy ground just doesn't pass through his mind. When I wrote in an email to him that Feynman had admitted that the problem is still there, he literally answered, "I am afraid even Feynman is out of date" and "There's been just a tiny little bit of progress over the ensuing decades of the 70s, 80s, 90s, 00s, and 10s." You see, these particle physicists are just wired differently. They think that Nature's problems become resolved by acclamation if they ape each other's arrogance long enough.

Folks like Fabiola Giannotti or Joe Incandela, the speakers of ATLAS and CMS, are among the relatively innocent ones. When Incandela said that the discovery was like finding a grain of sand in a swimming pool full of sand, he was aware of the boldness of such a claim, his face expressing, "Please, believe us!"

Much more shrewd was Rolf-Dieter Heuer with his well-planned joke that he had forgotten the name of the particle they had discovered. He is a master in making the public believe wrong things without lying explicitly. The jewel in the crown was his invention of the notorious hair-splitting between *the* Higgs boson and *a* Higgs boson, the *standard model* Higgs boson, and other (all sorts of non-standard, or what?) Higgs bosons. And Heuer did not lift a finger when the false news (also by "official standards," not mine!) that *the Higgs* particle had been discovered spread around the world in 2012. I think he just does his job, as sport functionaries and pharmaceutical lobbyists do. But he shouldn't be treated as someone with the credibility and reputation of a scientist.

Bookmakers and Buckmakers

There are many, too, who have some reputation in other fields of physics, who, though not having a clue about the experiment themselves, jump in and make a buck with their books which don't do more than parroting. The most insolent example is probably Lisa Randall, string-shooting star at Harvard and darling of science journalists. The sparsely covered 46 pages of her book contain just about half the characters of the Wikipedia entry on the Higgs. It takes some brashness to sell that. About the content, she had almost nothing to say in a recent interview.[89]

Randall seems such a sympathetic person as long as she talks about freshman physics or things like education, but when the interviewer asked the right questions (What did the LHC do for *your* theories? Are we going to reach some strange point where our ability to test these theories is limited? Isn't the future of your field so abstruse?"), she becomes increasingly nervous while pussyfooting around questions, and is downright lying[90] when she says that money limits the testability of modern theories, rather than string theory's continuous failure to predict anything. If the interviewer had asked her about the detector components that delivered the evidence for the thing she hyped, she would probably just have had her mouth agape. Another parroter coaxed the 89-year-old Leon Lederman to appear as the first (!) author of a book, *Beyond the God Particle*, trying a rehash of Lederman's 20 year old bestseller.[1]

Sean Carroll from the California Institute of Technology represents the cosmologists' chumming up with particle physics, talking about useless theoretical fancy while praising the standard models. He tried to crop his share of the Higgsteria with a book, *The Particle at*

---

[1] The title *God particle* should originally be *Goddamn particle* – meanwhile, a damn old story.

the End of the Universe, as if he wanted to say, "Hey, we cosmologists have also hyped a lot of nonsense, don't forget that." His presentation is more worked out than Randall's, it is ironic that in a way Carroll has collected all the information about the Higgs that is available – nevertheless it is the kind of rehash thousands of physicists could do, once they had read (and come to believe) the gospel of the standard model. Carroll, like most, then wanders off into wishful thinking about possible SUSY particles ("The hunt is on for yet more symmetries"),[91] a stuff that would help cosmologists if it is declared to be the long sought-after dark matter (probably they just don't understand gravity).[1] In the end, people like Carroll are not honest, they are fake experts. Their own scientific activity should reveal to them in the first place what it means to understand, in contrast to repeating sciolism. And if asked to talk about something you don't understand (because nobody understands what the heck the Higgs has to do with reality), you had better shut up. But the smart science presenters want to keep the semblance that physics has understood the world, with their posturing as untouchable wizards which would upset any of the deep thinkers of the past, who, in Carroll's place, would tell us that we haven't advanced at all since 1930.

Metabaloney

Most of the documentary stuff about the Higgs is about uninteresting meta-information that you can impart even if you don't understand anything of physics. The LHC has a circumference of 27 kilometers, it cost 9 billion dollars, its energy bill is like that of a major city, the protons cross the Swiss-French border, it is being paid for by 88 countries all over the world, the beam size is less than a needle and so on.[92] A pleasant read for any pinhead.

The underground beam, with the energy of a train, is 100 meters below ground, there are more than 5 million wires, it runs in an evac-

---

[1] Even the SUSY critic Peter Woit doesn't do other than parrot: "What they are doing is dumping a lot of energy into a small location, exciting the vacuum and telling something about its structure." (http://www.youtube.com/watch?v=irWoA_pEbQk).

uated tube, and the temperature there is "one of the coldest points in the universe" (liquid helium technology from 1911, it's quantum optics instead that holds the low temperature records.) All this is technology, not science. And psychology, rather than science it becomes when one tells a heart-rending story about the scientists involved in the experiment, as the *New York Times* did.[93] It is interesting to read about emotions and tears, but it has nothing to do with their physics output, which is absent.

Almost every documentary about CERN mentions the big bang that presumably happened some 13.7 billion years ago, but this is again meta- if not dis-information. This figure is the product of the Hubble telescope, you particle guys contributed nothing! But that doesn't prevent them from telling the fairy tale of the big bang being simulated in Geneva (actually, we are unable to simulate the interior of a common star like the sun). And it is easy to mock nuts who predicted the end of the world when the LHC was started up[94]

None of these documentaries say a word about the crucial factors of the discovery, on triggering, detector components, their material physics and artifacts, the energy calibration of the detectors on which everything depends, on the computer code which eliminates the background multiply stronger than the signal, and so on. Nothing. Why? Simply because none of the fake experts had anything to say about those topics.

Higgs Infotainment

But even when people are supposed to be not totally ignorant, like Brian Cox, a particle physicist who has become a kind of a science rock star, it is just some secondary information they offer you. To be able to repeat without stumbling is enough to be seen as a great physicist of our time (incidentally, the distinction between scientist and science seller has become invisible). The point is that guys like Cox sometimes behave like reasonable physicists, thus you don't realize immediately that they are particle physicists. I might like Cox's mockery of homeopathic medicine, or I even might agree with him in stressing how important physics education is for the economy (though that probably served just to hide the fact that particle physics has no useful output at all). It is also ok to publicize physics by slamming astrology, but to call it nonsense doesn't prove that high energy physics makes sense. Cox praises the "most complex machine ever built" which found the "genuinely fundamental" Higgs particle, according to him "one of the most important discoveries in the history of science, on equal footing with the electron," and he claims physics is in "a golden age."[95]

Verbiage. It's an age in which science is close to politics, and this is another sign of sickness. In response to the July 4 news, UK Prime Minister David Cameron immediately jumped on the bandwagon, prompting a wry comment by a BBC moderator who asked Brian Cox: "Do you think he had any idea of what he was congratulating the science community about?"[96] This just triggered a fawning statement from Cox about the effectiveness of Britain's science budget. What a rotten business physics has become.

Hilarious language hides what are just technical superlatives. But that's how physics popularizing works today. Combine that ever-smiling babyface with an emotional tremble in your voice, plus a smartass[1] historical ignorance and sufficient nebulosity in your pseudo-scientific talk, and the BBC will call you "the best person to explain the Higgs boson."[97]

Not Their Fault

It is tempting to blame journalists for their sometimes starry-eyed attitude and for letting the rigmarole off too uncritically. But they don't have a real choice in today's monoculture of scientists aping each other's blurry metaphors. Sometimes you distinctly feel the interviewer's uneasiness, but to bring the babble to the point is like nailing jelly on the wall. Expressing any serious doubt would cause a

---

[1] The Higgs gives mass to everything. Which sounds very bizarre, except that we now know that it is correct. According to Cox, it is "the best example of the unreasonable effectiveness of mathematics": http://www.youtube.com/watch?v=BeTjPsnzH-c

shitstorm about how embarrassingly ignorant the journalist's question was. Some occasional intelligent comments, such as Brian Cox being asked in the BBC interview mentioned above: "Doesn't it worry you that there is a total lack of knowledge in the Commons?", grasp at nothing. They aren't even worried by their own lack of knowledge.[98] Thus even the intelligent journalists give up, thinking, "Well I don't understand but that is not my business, I did my best to squeeze out the most understandable statements from that guy."

While celebrating the hype, most are cautious enough to foresee the consequences and provide against the opinion "well the LHC has bagged the Higgs, now we pack up and go home." "The great discovery," they add thoughtfully, has to be followed by "even greater efforts" of investigating it, notwithstanding nobody knows how, what and why. But all this is surely exciting (the word most used in describing this boring period). Jim Al-Khahili, another expert in pompous talking, said that "2012 promises to be a truly historic year for physics"[99] (with an historically hilarious comparison to 1905, Einstein's "miraculous year") and "we don't know what the masses of all these particles are" (true), but the Higgs might be the "key to telling us why these particles are as they are."[100] Al-Khahili continues with some soft babble about the Higgs being a possible "doorway maybe to other particles" (goodness, still more of a mess!), he is confident that it allows us "to tie down other uncertainties" (whatever that means), "looking in the right direction for other particles" (i.e. with blinkers) and being "on the threshold of another revolution." Yeah, the giant nonsense is about to crash. Sorry Jim, but that's not what one seeks as a physicist interested in fundamental questions. We need an *explanation* or, even better, a *calculation* with *testable* results, not a key, clue, signpost, direction or doorway. Dirac and Einstein would not have given a shit for that.

The Higgs Fake

"We seem to be very close to perhaps discovering"[101]

CERN knows how to do publicity. It's worth watching their infotainment-style[I] video collage of brief statements, including some Nobel laureates, entitled "The Higgs for Me."[102] Gerardus 't Hooft appears for three seconds, probably just to let this fairly critical mind not spoil the enthusiasm in the subsequent statement of David Gross, who "much more importantly" expects "new discoveries which give us clues," intending his pet supersymmetry. (The LHC found nothing that would give him a clue.)[II] Then Jerome Friedmann, Nobel prize winner of 1990, thoughtfully declares,

If the Higgs is discovered, it will be a great triumph for the standard model, if the Higgs is not discovered, it is practically certain that there is something in Nature that is equally interesting, maybe even more interesting that would create the symmetry breaking required by the standard model. And why do I say it's required? Because the standard model is so good.[103]

Savor this statement. His logic is so circular. Is there any space left for possible failure? Can Friedmann imagine an outcome that disproves the standard model? Sorry, but according to Popper, that plainly demonstrates that he got the Nobel for fiddling around with something that is unscientific.

Murray Gell-Mann, the 1968 laureate, has nothing to say about the experiment, but never misses the possibility to insinuate that he also contributed to an important discovery: "The mechanism that so

---

[I] Just have a look at crap like http://www.youtube.com/watch?v=ZgAWstsYxOQ&list=PLAk-9e5KQYEoxORIO9S4oYcFR5H4TwB7U and form your own opinion. Is it a coincidence that CERN more and more communicates its scientific message by cute, conforming greenhorns?

[II] After a talk in Munich in 2011 he responded to my question about if the LHC does not find a SUSY particle, "then we are in deep trouble." But they always gloss over any experimental failure and talk their way out.

many people proposed – Peter Higgs among many others [pause, smug smirk] – must be responsible for generating masses."

The ever-present CERN mascot John Ellis, as usual, expects the unknown to come next (you can hardly do wrong). And as usual, he doesn't understand that expecting the unexpected is not a scientific accomplishment, but a concession of ignorance if you have nothing else on your agenda.

Then four experimentalists appear on screen, all with the same smile of a just nursed baby, telling us that they were hunting the Higgs for twenty years, thus, whatever they found, it had to be the Higgs or some Higgs, and how exciting this is. No wonder it's exciting *them*, that indeed is what the spiritual glance in their eyes conveys. One of them tellingly declares, "I think I wouldn't spend twenty years of my life if didn't believe that there is a Higgs there" – replace "Higgs" with "God" and you have exactly the kind of statement that Richard Dawkins uses to mock them in public when confronted with religious nuts. And, as if this wasn't enough, the video (remember, an official CERN release) finishes with a choir singing a gospel, "Glorious Higgs".[1] What kind of shit are we being served here?

CERN, with its cultural products "LHC Rap"[104] and "Particle Fever", undoubtedly inspires the music and film industry. But today's physics no longer has anything to do with the great accomplishments of the late nineteenth and early twentieth centuries. It is time to stop. Time to reflect. Time to dump a big science enterprise that has grown to absurd complication in every sense, has swept under the rug the important problems, has developed nothing useful and impedes any true progress in understanding the laws of Nature.

---

[1] CERN physicists probably missed the irony when they took the music from the song (http://www.youtube.com/watch?v=1QW85kfakJc) "glorious mud". That perfectly describes the gamma-gamma-channel of the Higgs evidence.

The Higgs Fake

CHAPTER 13

AGAINST BELIEF: WHAT WE NEED TO DO

...AND WHAT YOU SHOULD ASK THESE GUYS

When you discover that you are riding a dead horse, the best strategy is to dismount. – Wisdom of the Dakota Indians

If there is a comment I expect, it will be that my critique is cheap and superficial. But how do you like it? Should I analyze and comment in detail thousands of papers that have appeared in the field, and hope to be published? The problem is that the underlying hypotheses of particle physics have grown to a complexity that cannot be overseen any more. Without a detailed study, without years of experience, nobody can claim to get an idea of the field, let alone understand it. In other words: before you criticize the business, get employed in it first. This is, sadly, a form of social nonfalsifiability, not too far from esoteric, of which the understanding is also granted only to those who can believe. What would be needed instead is a whistleblower program that encourages reflective people in the community to state their opinions. The groupthink dropouts should not lose their jobs, or even better, they should be assured to get a new one. That's how you have to fight a sect.

Alexander Unzicker

Asked about the future progress of physics, today's particle physicists advocate the moronic activity of collecting still more data at higher energies, with the bizarre hope that "clues" for a simplification will suddenly drop out of the ever increasing complication. But there are already too many free parameters floating around for anybody to stop the inflation of this currency of fundamental physics. Continuing collider physics at higher energies just leads physics deeper into the mess. It's like printing money.

Large communities do not necessarily generate responsible people. In today's busy science environment there is not a single position that would allow anyone to think about the fundamental questions for the time Newton or Kepler needed. We probably should pay physicists like monks – financing frugal lives of smart people who dedicate themselves to science, without bothering them with futile public relations activities.

Break It to Pieces

My soup is oversalted: Should I be disallowed to call it oversalted until I can cook it myself? – Gotthold Ephraim Lessing, German writer

"What's your alternative?" is a rebuke I have heard several times. It truly means: "Come up with an alternative that reanalyzes all the existing data (done by tens of thousands of physicists), prove that it yields better results and convince the community of it." The problem is that radical alternatives are not appreciated very much. It is easy to find a consensus for extending an existing theory (which increases the number of free parameters), but woe betide anyone scrapping an established concept to which people have dedicated years of research. Thus it is utterly impossible to open a discussion on, say, the soundness of the quark concept. It's a muddle. No genius whatsoever could prove his hypothetically correct theory of everything because all experimental outcome is formulated in the messy, nonsensical picture of the standard model, in terms of branching ratios of W bosons, bottom quarks, and all that hogwash. Completely useless for any possibly reasonable analysis. So what to do? I will be blamed anyway for not offering an alternative theory, but here is my proposal for an alternative treatment of the data. In general, a theory-independent representation of the data is needed. This is easier said than done, because a lot of triggering and filtering options have been chosen according to the "needs" of the standard model. Thus particle physicists will probably argue this is impossible in principle, because they for generations they have been used to their theoretical-experimental blend that now seems inseparable. But it is.

The first step would be to reorganize and divide the detector groups (ATLAS, CMS, Alice, LHCb) into four to six completely separate and independent sets of scientists. Each subgroup should be responsible for a different stage of evaluation of the data. At the first, basic level, material science experts should deal with the detectors (that group, of course, could be split up according to the respective types of detectors), and should calibrate them, which means determining the detectors' response to a photons, electrons, protons, neutrons, muons of *known* energy, simply to know in advance how your detector responds before you measure or even interpret the outcome. And this calibration *has to be published* before anything else happens. It is of utmost importance that the people involved don't have the slightest interest whether a Higgs comes out out of the final analysis or not.

What Should Be There?

The scientist is free, and must be free to ask any question, to doubt any assertion, to seek for any evidence, to correct any errors. — Robert Oppenheimer

To repeat: the data that matters is how much energy at a given time is deposited at a given detector location. Period. And this must be open to anyone's interpretation, and everyone has to have the possibility to analyze it. It might be a considerable piece of work, yes, but the era of preferred knowledge of the collaborations must end.

If there is a selection of data (triggering) the criteria of how to do this selection must be not only open, but there should be the possibility for any researcher to get a focus on what he considers important. It's rather like if you have a big telescope and must decide where to point it (the large synoptic space telescope, allowing for internet access, does exactly this).

And of course, even if the data selection is determined by individual choice, the access has to be open to anybody. Once the more elementary processes of data evaluation like calibration are reliably done, indeed many people may welcome cleansed data on a second stage, such as what energy the detected particle had (in contrast to the first stage of the pure detector response). And again another group, making use of their calibration, should make public this data. I don't object if then, at a more sophisticated level so to speak, another research group analyzes the data in terms of W bosons and bottom quarks. But you must not to impose that stuff on everybody. It is essential that a skeptic can enter at the level *he likes* and scrutinize the analysis there. And it is important that there are different kinds of skeptics at all levels.

To be fair, there are tiny little programs that deal with the issue of making data publicly less inscrutable. Both ATLAS and CMS offer sites where you can download collision data,[105] such as a selection of events with two muons, where the energy and flying direction is given (a characteristic which is relatively easy to determine from the curvature of the particle path in a magnetic field). If this kind of program expands (the head of the program, as far as I can tell, is seriously interested in making people participate), it would be a first step towards the right direction. However, at the moment, it is little more than a kids' playground. When you click on the site, the first thing you are given is a warning that the data is not suitable for scientific purposes, not authorized, not verified in this and that. Are they afraid that a schoolboy may disprove their analysis by having a quick look at an Excel file? (Well, it happened to two economics celebrities from Harvard.) Anyway, that program has still to prove whether it is designed for young students being brainwashed as early as possible by some unthinking exercises in how to "identify" W bosons, or whether it is the first step towards a paradigm shift in how scientific data are processed. My hope is, even if it is designed for the former, it will result in the latter.

Freedom is the first-born daughter of science. – Thomas Jefferson

The Higgs Fake

Collective Activity Needed? Yes, but...

You might have gathered that I do not deny the necessity of a collaborative effort when analyzing a huge amount of data. The question is how such a collaboration is organized. Wikipedia is also a collective project, far beyond the capacities of an individual, but it does not work in a manner that 5000 experts agree what needs to be in an article and then publish it. What we need is indeed a Wikipedia-like culture of dealing with scientific data: open, transparent, neutral point of view, room for minority opinions. That would be my dream: you order your experimental setting via mouse click, the particle collision is registered, stored in standardized file formats, and then you can do what you want – get it pre-analyzed by other people's code by another mouse click, or write your own code that picks out a new aspect that nobody has addressed before.

Yes, particle physicists, we are still very far from that and you are free to complain about my lack of knowledge in complicated details and to mock my supposedly somehow naïve visions. But at the same time I feel free to mock CERN for not making use of the only useful thing it has produced, the World Wide Web. You may blame me for the incompleteness or partial infeasibility of my proposal. But it is not my business to give a detailed program for how to transform something that has grown beyond any tractable size back to a reasonable experiment, it's your business. And don't lament that you have no resources and you need to apply for extra funding before governments eventually grant money (that would be an excellent strategy to delay a reasonable use of the data until the cows come home). Start here and now. You have the computers, you have thousands of people involved in the experiment, start explaining it.

And please don't resort to stupid counterarguments. One particle physicist, in all seriousness, argued that only the experts have the type of detailed knowledge needed to conduct a sound analysis. Thus making the raw data (technically called minimum bias data) public, due to the danger of false conclusions, would be "counterproductive." Not only do I think this is crap, but I do believe that people stating such contrived nonsense constitute independent evidence that their business is rotten. It's not *your* detector, guys, it's the taxpayer's.

Thus don't continue claiming that only you understand it and only you can make sense out of the data, or similar justifications for your monopoly. Explain instead what needs to be explained, document it, make tutorials, make everybody the hell understand what you are doing. The world wants the unspoiled data, so get it on the table. It may be difficult to create transparent access to the data, but there is no way around it, so suck it up.

The data is the only objective, unprejudiced, opinion-free thing we have. And don't bemoan that it is too much to store. This is your problem, guys. There are dozens of particle accelerators in history of which we *could* store the complete data. Go and do that. We want to know what came out of CERN, Tevatron, DESY, SLAC... just to name the most prominent ones. Is the data stored in old file formats? Go and get it from the tapes. Doesn't exist anymore? Go and rebuild the old machine (it will cost the air conditioning bill of the CERN control room). We want to know how all came about. Don't believe it? The public wouldn't be interested? No. People are uninterested in impenetrable garbage. It just serves your purposes that the stuff is perceived as something too difficult to grasp for the average physicist. Once the subject becomes intelligible, humankind would love to see, much like in astronomy, how the elementary building blocks of Nature behave.

There is no way around transparency. The internet has the power to revolutionize science, and if there is a field that desperately needs it, it is particle physics. The data has to be open, accessible, controllable, and supported. Besides all the other good reasons, I think there is a particularly convincing one: At the moment, to give a careful estimate, roughly one percent of the planet's population may have the chance to dedicate themselves to real science, because the living conditions make that tiny selection. Besides the things I had to say about science, there, too, a change has to come.

The Transparency Agenda for (Particle) Physics

1. World Scientific Heritage

Every experiment that provides evidence for a fundamental concept should be continuously repeated, possibly also with a copy of the original apparatus.

2. Completeness

The goal must be to maintain the complete data of an experiment. If there is a selection (filtering, triggering), it must be a representative set of the complete data.

3. Backup

All data must be stored permanently and as completely as possible after every stage of the analysis.

4. Documentation

Every step of the data processing should be justified and documented, every step of data selection should be verifiable by open-source programs.

5. Data Reduction

Every step – elimination of artifacts, calibration, corrections, removal of background, extraction, comparison with simulations etc. – should be performed by modules and independently organized groups.

6. Openness

All data in raw and processed form must be publicly accessible, using standardized file formats, to guarantee a model-independent analysis.

7. Calculability

All evaluation steps must be repeatable, ideally with remote access, while parameters can be varied by any user.

8. Reproducibility

The whole experiment must be as transparent as possible, including using photo and video documentation.

9. Metadata

All metadata, even if considered irrelevant by the mainstream paradigm, should be recorded, at least the time and meteorological conditions of the environment.

10. Support

External reproduction of the experiment must be supported by scientists directly involved with the experiment.

What You Need To Ask These Guys

The standard model is complicated beyond any credibility, it does not address a single fundamental problem of physics, it is a textbook example of a Kuhnian crisis, its experimental techniques make it increasingly more likely that researchers fool themselves, knowledge in the community mostly consists of parroting expert opinions, and the experiments are practically not repeatable and completely nontransparent. The standard model is bad science. Nobody has to be afraid of stating that. To debunk the omnipresent sophisticated propaganda, I recommend asking questions along the following lines (don't say that they are from Unzicker's book).

1. Q: A free parameter is a number which is fitted to the measurements to describe the outcome. Can you tell me how many free parameters the standard model contains?
   - *(17 to 50):* Is a high number of parameters a good or a bad sign for the validity of a physical theory? *(obviously bad)*
   - Can you give an account of how the number of free parameters developed in the last 80 (50, 30) years? *(steady increase)*

2. Q: The standard model has so many free parameters. Doesn't it remind you of the medieval epicycle model?
   - *No:* So do you think we have to put up with 20 unexplained numbers forever?

- *Yes, but we have no better choice but to continue to work.* So Copernicus, Kepler and Galileo continued to work on the epicycle model?
- *It is the only model we have.* Would you fly across the Pyrenees if the only map you have is of the Alps?

3. Q: Accelerated electrical charges radiate, but physics doesn't have a formula that says precisely how (a problem that goes back to classical electrodynamics). How can you know the energy radiated away at huge accelerations (during the crash)?
    - *It is (theoretically) known:* So Richard Feynman is wrong in stating that the problem isn't solved yet?
    - *It is solved by later theoretical work.* This would be a great accomplishment. So who got the Nobel for it?
    - *We know the radiated energy by calibration.* How did you do that, given that the experiment has unprecedented energy?

4. Q: As recent results from quantum optics show, the proton radius is 4 percent smaller now than was assumed earlier. Doesn't that screw up the analysis of the behavior of colliding protons?
    - *The size is irrelevant.* So nothing would change if the proton had the size of a pea?
    - *That holds for quantum optics:* Doesn't physics rely on checks across its subfields?

5. Q: Can the standard model predict the masses of elementary particles?
    - *Yes (nonsense):* So if it can calculate masses, how does the physical unit kilograms appear? (impossible by any known combination of natural constants)

- *No, but it is not important:* Is mass just an irrelevant property in physics or would it be desirable if the standard model could calculate?

6. Q: Is gravity important for particle physics?
   - *No (stupid):* Aren't the concepts of mass and gravity deeply related? (By Einstein's principle of equivalence)
   - *Yes, at the Planck scale.* Can that ever be tested? (*No.*)

7. Q: The relation of the electric to the gravitational force is a huge number, 10 to the 39th. Can the standard model explain this?
   - *No (only possible answer)*: Shouldn't fundamental physics have answers to such questions?

8. Q: It is said that the Higgs mechanism explains masses. Does that mean that one can now calculate a number like the mass ratio of proton to electron, like 1836.15?
   - *No:* Why not? Wouldn't it be desirable to have a theory that can?
   - *Yes, in principle (babble about lattice quantum field theory)*: So what is the accuracy? (ridiculous)

9. Q: The standard model of weak and electromagnetic interactions seems complete now. Can the standard model calculate the fine structure constant? (1/137.035999...)
   - *No (only possible answer):* Why not?

10. Q: Have theories in the history of science developed in a steady way, or sometimes abruptly?

11. Q: Aren't many ad-hoc assumptions and unexplained numbers a sign of a Kuhnian crisis in the standard model of particle physics?

12. Q: It is said that the standard model of particle physics is precisely tested. Which is the best of all tests, in terms of a theoretically predicted number, to agree with a measured value? Please give an accuracy in percent.
    ◦ *The mass of the top quark:* Was that predicted? So why did physicists look for it from 1977 to 1995 in much lower mass ranges?
    ◦ *The Weinberg angle/ the weak coupling constant/ The masses of W and Z bosons (to about 1.7 percent):* Compared with general relativity or quantum electrodynamics, isn't that poor?
    ◦ *Any better accuracy:* Where is it published?

13. Q: Just one of hundreds of billions of photon pairs is said to originate from a Higgs decay. How can one unambiguously separate such a tiny signal from such a huge background, given the complexity of the experiment?
    ◦ *The background is well-known:* How can it be, if the energy region is unprecedentedly high?

14. Q: How can one distinguish if a particle shower is generated by a neutron or a neutrino?
    ◦ *By theoretical models only:* So how can one speak of an experimental evidence for neutral currents, which are based on that distinction?

15. Q: Does the Higgs discovery mean that the standard model is correct?
    ◦ *Yes:* Then shouldn't it be able to give predictions for masses? Or is there anything that cannot be understood as a matter of principle?
    ◦ *It is not complete (or similar babble).* By complete do you mean a theory that has more free parameters or less?

o   *Future experiments might give clues.* Of course, but is there any new information in this statement?

16. Q: Among 3000 people, there must be dissenting opinions about the data analysis. Why are these opinions not published?

17. Q: If I wanted to know every detail of the experiment and its evaluation, how many people would it require to have that complete expertise?

18. Q: If our civilization was destroyed and just the computers and printed publications remained for posterity, would it be possible to rebuild the LCH experiment (or the Higgs discovery) on this basis?
    o   *Yes:* And what if there were just the printed publications?

19. Q: Can anybody outside the collaborations verify the analysis, with publicly available data?
    o   *Yes:* So where is the public access to the raw data?
    o   *No:* Wouldn't it be a sign of transparency to disclose the data analysis?
    o   *Data disclosure would risk false analyses by non-experts:* Are there any negative experiences in astrophysics, where the data is public?
    o   *Only the experts know the detector:* Isn't it part of the scientific method to make such knowledge public?

20. Q: Maxwell's electromagnetism is tested by cell phones, Einstein's general relativity by GPS receivers, and quantum mechanics by digital cameras. Can one expect a similar application for the Higgs, for W bosons, for top quarks?
    o   *No (honest):* Haven't the big discoveries always transformed into technology?

- *(babble):* Can you be more concrete?
- *Yes:* Curious to hear that!

21. Q: What do you hope to find next?
    - *Supersymmetric particles, dark matter particles, strings...* How should this manifest? Can you give a numeric value for the prediction?

22. Q: You say that the LHC may discover something unexpected, which is a sign of great discoveries. But doesn't such an aspiration imply that the surprise is limited?

23. Q: Can you give an example of a *new* LHC result, which does not contradict earlier experiments but that would suddenly disprove the standard model?
    - *Look, we know that the standard model is only an approximation (like Newtonian) physics:* But Newtonian physics wasn't known to be an approximation, until it was superseded by Einstein's theory. How do you know in advance?
    - *If an experiment proves quantum mechanics wrong, that would mean the standard model is wrong, too:* I said an earlier experiment. Please refer to something after 1960.

24. Q: It is clear that today's theories are untestable by technology, largely a consequence of the size of our planet or similar unalterable conditions. Do you think that the supreme insight a civilization may achieve in the universe is limited by such environmental conditions or by its intelligence?

LITERATURE

Barbour, Julian. *The Discovery of Dynamics*, Oxford University Press, 2001.
Collins, Harry. *Gravity's Shadow*. Univ. Chicago Press, 2004.
Einstein, Albert. *The World as I see it.* Open Road, 2011.
De Solla Price, Derek. *Little Science, Big Science and Beyond.* Columbia Univ. Press, 1986.
Feyerabend, Paul. *Against Method.* Verso, 1975.
Feynman, Richard. *Lectures vol. II.* Basic Books, 2011, chap. 27+28.
Feynman, Richard. *QED.* Princeton University Press, 1988.
Galison, Peter. *How Experiments End.* Chicago University Press, 1987.
Gleick, James. *Genius.* Pantheon Books, New York, 1992.
Heisenberg, Werner. *Physics and Beyond.* Harper & Row. 1972.
Horgan, John. *The End of Science.* Addison-Wesley 1996.
Jones, Sheilla. *The Quantum Ten*, Oxford University Press, 2008.
Kuhn, Thomas. *The Structure of Scientific Revolutions*, Univ. Chicago Press, 1996.
Landau, Lev. *Theoretical Physics* II. Pergamon Press, 1975.
Lederman, Leon.*The God particle.* Houghton Mifflin, 2006. (12,13)
Lindley, David. *The End of Physics.* Basic Books, 1994.
Pickering, Andrew. *Constructing Quarks.* Univ. Chicago Press, 1984.
Popper, Karl. The Logic of Scientific Discovery. Routledge, 2002.
Rosenthal-Schneider, Ilse. Begegnungen mit Einstein, von Laue und Planck (German). Vieweg, 1988.
Schrödinger, Erwin. *My View of the World.* Cambridge Univ. Press, 2008.
Segrè, Emilio. *From X-Rays to Quarks.* Univ. California Berkeley, 1980.
Unzicker, Alexander, Jones, Sheilla. *Bankrupting Physics.* Palgrave 2013.
Weinberg, Alvin. *Reflections on Big Science.* MIT Press, 1969.
Taubes, Gary. *Nobel Dreams.* Random House, 1987.

The Higgs Fake

Index

Accelerated charges 85, 87, 104, 119
Al-Khahili 131
Alvarez 35, 81
ATLAS 56, 64, 67, 70, 107, 111, 117f, 121, 126, 136, 137
Atomic bomb 34, 37, 44, 69
Barbour 27, 33
Big science 37, 55, 60, 62, 94, 111, 133
Bohr 16, 20, 25, 35, 61, 69, 73, 75, 80, 94, 111
Bottom quark 42, 101, 105, 113, 122, 135, 137
Carroll 127, 128
CERN 5, 9, 14, 18, 28, 29, 37, 38f, 41, 43, 49, 51, 56ff, 62, 63, 67, 70, 72, 93, 97f, 105ff, 109ff, 118, 120, 129, 132f, 138f
Charm quark 101
CMS 43, 56, 64, 67, 70, 107, 118, 121, 126, 136, 137
Constants 16, 22, 28f, 46, 74, 142
Cox 129ff
de Broglie 7, 35
DESY 17, 72, 139
Dirac 5, 7, 12, 20, 23, 28f, 35, 37, 51, 66, 74f, 77, 80, 87, 90, 94, 104, 131
DORIS 98ff
Einstein 5, 7, 12, 16, 20f, 24f, 27, 29, 34ff, 46, 51f, 61ff, 66, 69, 73, 77, 80, 82, 90, 94, 95, 103f, 111, 115, 122, 131, 143, 145f
Epicycles 33
Faraday 62, 66, 98, 110f
Fermi 35, 81, 100, 110
Feynman 21, 27, 28, 60, 75, 76ff, 86, 89f, 115, 126, 142
Gell-Mann 35, 86, 89f, 103, 104, 116, 132
Geocentric model 5, 18
Glashow 35, 92, 98, 108
Gross 35, 102, 132
Heisenberg 7, 26, 36, 60f, 73, 75f, 119
Heuer 43, 57, 67, 120, 126

Higgs boson..............................11, 39, 42f, 66, 69, 118,f, 126, 130
Higgs mechanism 23, 115ff, 122, 123, 143
Higgs particle........................... 5, 23, 116, 120, 121, 125, 126, 130
Isospin 12, 25, 85f, 88, 104, 124
J/psi particle 95ff, 100f
Large Hadron Collider................................... 9, 21, 55, 58, 62f
Lederman 35, 72, 100, 101, 109, 127
Lindley 6, 27, 84, 102, 116
Mach 27, 37, 63, 77, 81, 84, 92, 97, 100f, 107f, 119, 130, 139
Maxwell 12, 15, 29, 35, 94, 111, 145
Methodology 65, 66, 93, 107, 122
Muon 11, 45, 47f, 72, 78f, 96, 99, 120, 122, 136, 137
Neutral currents....................................38, 72, 91, 92, 93, 144
Neutrino 11, 25, 26, 45ff, 60, 69, 72, 78ff, 91, 92, 107f, 123, 144
Newton..................12, 15, 17f, 22, 34, 35, 94, 98, 111, 135, 146
Pauli 20, 22, 73ff, 78, 89, 94, 101
Pickering 6, 24, 32, 49ff, 65, 91, 92, 96, 98, 103
Planck 16, 35, 36, 53, 97, 143
Popper 14, 39, 83, 132
Randall..127f
Reproducibility...66ff
Richter 35, 97, 99, 108
Riddle of mass..23, 122
Rubbia 35, 55, 106ff
Rutherford 60, 83, 86, 110
Schrödinger.....................5, 7, 12, 20, 26, 36, 37, 51, 61, 73f, 76f, 80, 84
Segrè 88, 100, 123
SLAC 91, 97, 139
SPEAR 96, 98f
Spin 24ff, 29, 90, 101, 116
Standard model 6, 8, 9, 16ff, 22ff, 29, 31ff, 37, 39, 45ff, 52ff, 58f, 62, 69, 79f, 88, 96, 105, 112ff, 116, 122, 123ff, 132, 135f, 141ff, 146
Strangeness 25, 85, 86, 87, 88, 98, 104, 124
Symmetries 25, 33, 102, 128
Symmetry breaking....................................... 27, 117, 132
Taleb 13, 19, 32, 38, 44, 53, 55, 70
Tauon..11
Tevatron... 68, 72, 113, 117, 139
Top quark 42f, 43, 49, 69, 71, 113ff, 117, 119, 144f
Trigger 42, 49, 65, 70, 112, 129f, 136, 140

W boson 42f, 46ff, 67, 69, 71f, 107f, 113, 118f, 135, 137
Wegener 36
Weinberg 32, 35, 72, 92, 99, 105, 108, 112, 116, 123
Z boson 42, 45, 47, 67, 107, 119, 144

Endnotes

[1] About fundamental constants, Rosenthal-Schneider, p. 24ff.
[2] Landau II, par. 75.
[3] Feynman (Lectures II), chap. 28.
[4] Dürr et al. Science 322 (2008), 1224–1227.
[5] Rith et al., The mystery of the nucleon spin, Scientific American, July 1999, p.58.
[6] Pickering, p. 12.
[7] Schrödinger, Die Natur und die Griechen, p. 26.
[8] A more detailed account of this is found in arxiv.org/abs/0708.3518
[9] arxiv.org/abs/physics/0110060.
[10] The New York Review of Books, October 8, 1998, pp. 48-52.
[11] Barbour DoD, p. 141.
[12] Though with a change of paradigma, cfr. Kuhn p. 115.
[13] www.positiveatheism.org/hist/quotes/russell.htm.
[14] arxiv.org/abs/1306.0571.
[15] arxiv.org/abs/1203.0275.
[16] http://www-alt.gsi.de/documents/DOC-2003-Jun-25-4.pdf.
[17] http://www.nytimes.com/1984/06/25/us/physicists-may-have-tracked-last-quark-to-lair.html
[18] Pickering, p. 9.
[19] Pickering, p. 404.
[20] Albert Einstein. *Mein Weltbild*. Ullstein 2005, p. 118.
[21] Taleb, Antifragile, p. 419.
[22] http://en.wikipedia.org/wiki/Asch_conformity_experiments
[23] http://www.nobelprize.org/nobel_prizes/physics/laureates/2005/hansch-bio.html
[24] Pickering, A. 1981. "The Hunting of the Quark". Isis 72: 216–236.
[25] For example, http://profmattstrassler.com/2012/11/14/higgs-results-at-kyoto/
[26] Taubes, p. 90.
[27] http://rjlipton.wordpress.com/2013/03/07/checking-the-higgs-boson/
[28] https://twiki.cern.ch/twiki/bin/view/AtlasPublic/HiggsPublicResults shows just the final graphs, for example.
[29] Taubes, p. 87.
[30] Nature 474 (2011), p. 16f., www.nature.com/news/2011/110527/full/474016a.html.

31 Taleb, Antifragile, p. 395.
32 Kragh, p. 184f.
33 Kragh, p. 166f.
34 Kragh, p. 166f.
35 Feynman(QED), chap. 4, p. 149.
36 L. Wang, Arxiv.org/abs/0804.1779.
37 F. Dyson, Phys. Rev. 85 (1952), p. 631.
38 E.g. Kamiokande, Phys. Rev. Lett. 81 (1998), p. 1562ff, arXiv:hep-ex/9807003.
39 TAUP taup2011.mpp.mpg.de/?pg=Agenda
40 See J. Ralston, Arxiv.org/abs/1006.5255.
41 G.A. Miller, Phys.Rev.Lett. 99 (2007), p.112001.
42 D. Lindley, Nature 347 (1990), p. 698.
43 Pickering, p. 129, see p. 153 for the questionable methods.
44 Pickering, p. 208.
45 Pickering, p. 139.
46 Dirac, Nature 126 (1930), p. 605.
47 Gleick, p. 433.
48 Pickering, p. 147.
49 Pickering, p. 140.
50 Pickering, p. 143.
51 Pickering, p. 188: "The equivalence is not transparent to the untrained eye, but nor is it to the trained eye."
52 Pickering, p. 192.
53 Pickering, p. 244.
54 Pickering, 258.
55 Taubes, p. 59.
56 Pickering, p. 262.
57 Pickering, p. 266.
58 Pickering, p. 267.
59 See e.g. Pickering, pp. 288-293 on `optical rotations'.
60 Pickering, p. 282.
61 Segrè, p. 291.
62 Lindley, p. 119.
63 A. Pickering, Isis 72 (1981), p. 216–236.
64 Pickering, p. 111.
65 Taubes, p. 94.
66 Taubes, p. 7.
67 Taubes, p. 6.
68 Taubes, p. 78
69 Taubes, p. 89.
70 Taubes, p. 79.

The Higgs Fake

71 Taubes, p. 82.
72 Taubes, p. 85.
73 Taubes, p. 84.
74 Taubes, p. 90.
75 Taubes, p. 107.
76 http://www.youtube.com/watch?v=P-U2nGFqupI
77 Taubes, p. 78.
78 Taubes, p. 124.
79 Scientific American 9/1997, p. 58.
80 Tipton, Scientific American 9/1997, p. 59.
81 cf. Gleick. Genius.
82 Lindley, p. 172.
83 www.math.columbia.edu/~woit/wordpress/?p=4553.
84 Feynman II, chap.28, and Landau, par. 75 make clear that in general, huge accelerations are not understood.
85 R. Pohl et al., Nature 466 (2010), p. 213.
86 https://atlas.web.cern.ch/Atlas/GROUPS/PHYSICS/CONFNOTES/ATLAS-CONF-2012-093/
87 http://quotationsbook.com/quote/20120/#sthash.JSomkdRo.dpuf
88 www.youtube.com/watch?v=dfoq8kFC6xc
89 http://www.youtube.com/watch?v=yDHta5Fq93M
90 http://www.youtube.com/watch?v=yDHta5Fq93M 26:30.
91 4:00 min.
92 http://www.youtube.com/watch?v=BeTjPsnzH-c
93 http://www.nytimes.com/2013/03/05/science/chasing-the-higgs-boson-how-2-teams-of-rivals-at-CERN-searched-for-physics-most-elusive-particle.html?view=Oozing_Into_View
94 http://www.youtube.com/watch?v=r7BTqKeP6Ks 7:00 min.
95 http://www.youtube.com/watch?v=sSLGaUQRWHg
96 http://www.youtube.com/watch?v=BeTjPsnzH-c 3:50 min.
97 http://www.youtube.com/watch?v=yKz07k04D70
98 http://www.youtube.com/watch?v=BeTjPsnzH-c 4:30 min.
99 Horizon special, The Hunt for the Higgs, http://www.youtube.com/watch?v=af1FyQyTWT4
100 http://www.youtube.com/watch?v=Mq6VHedjIEI
101 www.youtube.com/watch?v=anhPnnyocxA, 2:30 min.
102 www.youtube.com/watch?v=anhPnnyocxA
103 www.youtube.com/watch?v=anhPnnyocxA, 0:47 min.
104 www.youtube.com/watch?v=f6aU-wFSqt0&list=TLNYsohA4AsQM
105 http://cms.web.cern.ch/content/cms-public-data-samples

Printed in Poland
by Amazon Fulfillment
Poland Sp. z o.o., Wrocław

92702402R00087